From RAINBOW to GUSTO:
Stealth and the Design
of the Lockheed Blackbird

From RAINBOW to GUSTO: Stealth and the Design of the Lockheed Blackbird

Paul A. Suhler

Ned Allen, Editor-in-Chief
Lockheed Martin Corporation
Palmdale, California

Published by
American Institute of Aeronautics and Astronautics, Inc.
1801 Alexander Bell Drive, Reston, VA 20191-4344

American Institute of Aeronautics and Astronautics, Inc., Reston, Virginia

1 2 3 4 5

Library of Congress Cataloging-in-Publication Data

Suhler, Paul A.
 From Rainbow to Gusto : stealth and the design of the Lockheed Blackbird / Paul A. Suhler ; Ned Allen, editor-in-chief.
     p. cm. -- (Library of flight)
  Includes bibliographical references and index.
  ISBN 978-1-60086-712-5
  1. SR-71 Blackbird (Jet reconnaissance plane) I. Allen, Ned. II. Title.
  UG1242.R4S84 2009
  623.74'67--dc22

2009030147

Copyright © 2009 by the American Institute of Aeronautics and Astronautics, Inc. All rights reserved. Printed in the United States of America. No part of this publication may be reproduced, distributed, or transmitted, in any form or by any means, or stored in a database or retrieval system, without the prior written permission of the publisher.

Data and information appearing in this book are for informational purposes only. AIAA is not responsible for any injury or damage resulting from use or reliance, nor does AIAA warrant that use or reliance will be free from privately owned rights.

Foreword

Deep in the Appalachian woods stood the still of cartoonist Andy Capp's mixed-culture moonshiner Big Barnsmell. The fires sparkled, smoke contrails threaded through the trees, and the odors permeated the forest. This was where "inside man" Barnsmell made Kickapoo Joy Juice for *Li'l Abner's* secret *Skonk Works*. During WWII Lockheed Aircraft Company was busily executing a secret contract to build the first US jet fighter, working day and night, under a rented circus tent as there was no room in the plant fully committed to assembling propeller-driven aircraft. To Lockheed's employees, the sounds, smells and welding flashes coming from that tent, the pungent odor from a neighboring plastics factory, against a backdrop of wartime suburban Los Angeles, recalled Barnsmell's "Skonk Works" and the phrase was adopted as their whimsical moniker for the place.[1] Eventually, "Skonk Works" became Skunk Works® and synonymous with Lockheed's "black" aircraft.[2]

When I first visited the headquarters of Lockheed Martin's Advanced Development Programs unit, the official name for the Skunk Works, I was flushed with excitement and not a little fear. I had been invited to make a presentation to its most senior engineers and program managers on a project we had been working in my own small company back in the rural mountains of northern Utah. To me, the famed Skunk Works was to aeronautics what Harvard and Oxford were to higher education ... the best place on earth for visionary work. I recall my walk down the long hallways, passing door after closed door with Diebold locks in lieu of keyholes and electronic keypads on the adjacent jams ... and no windows. No windows at all, not even in the few offices I was escorted into. That first impression was awe, a little fear that

---

[1] Irven H. Culver, a self-educated aviation engineer whose ingenuity earned him many of his field's highest honors, died Aug. 13, 1999 at 88. "One day during the war, Culver answered ... 'Skonk Works, inside man Culver speaking.' The call was from a Navy officer who 'laughed and asked me to repeat it while he put on a loudspeaker in his Washington office so everyone else could hear it,' Culver once said. Division chief Kelly Johnson did not laugh, however, and fired Culver. But Culver, whose antics got him fired 'at least twice a day,' survived the incident, as did the Skunk Works name ..." ('Culver obituary' LA Press, Sept 17, 1999).

[2] "Skunk Works" and the skunk logo are registered trademarks of Lockheed Martin Corporation.

I might inadvertently see some secret successor to the fabled Blackbird (the rumored Aurora? I mused ...) and be escorted away by security men to an endless interrogation.

Today, the stealth technologies, first put in the air by the Skunks with the U-2 and then greatly refined in the SR-71 have since been extensively perfected and, in spite of their sensitivity, diffused throughout the industry and successfully adopted by many: Northrop Grumman, Boeing, General Atomics, among others. But the real story in these events, so directly and forcefully recounted by Paul Suhler, lies with the strangeness of a past engineering culture accepting of so much schedule and technological ambition and with such tolerance for technical and personal risk as the U-2 and SR programs were. Today's aerospace engineering and development environment seems tame by comparison. So, Paul Suhler's history of those days holds a special place among the readers of the LIBRARY OF FLIGHT.

The LIBRARY OF FLIGHT is part of the growing portfolio of information services from the American Institute of Aeronautics and Astronautics. It augments the two existing book series of the Institute, the 'Progress' in aerospace series of technical monographs and the 'Education' series of textbooks, with the best of a growing variety of other topics in aerospace from aviation policy, to case studies, to studies of aerospace law, management, and beyond. RAINBOW and GUSTO: Stealth and the Design of the Lockheed Blackbird is a very welcome contributor to the series.

Ned Allen
Editor-in-Chief, LIBRARY OF FLIGHT
Bethesda Maryland
June 2009.

TABLE OF CONTENTS

PREFACE xiii
ACKNOWLEDGMENTS xv

CHAPTER 1   INTRODUCTION 1

"A Unique Opportunity" 2
Designing the U-2 5

CHAPTER 2   RAINBOW 11

Beginning the Search for Invisibility 13
Stealth in 1956 16
Mis-Proof of Concept 18
U-2 Measurements 19
Wallpaper 22
Trapeze 25
Wires 28
Security 30
Flight Testing 32
Westinghouse Experiments 33
RAINBOW Missions 35
The Ship Is Still Sinking 37
Passive ECM Committee 37
Flying Saucer 38
Scientific Engineering Institute 40

CHAPTER 3   RAINBOW PHASE II 45

Interception Study 45
New Reconnaissance System 46
Basic Stealth Techniques 48
Summing Up 50
Tracking and Interception Probabilities 53

Chapter 4 The Follow-On 55

B-2: A Reshaped U-2 55
GA #2 58
Project GUSTO 60
More RAINBOW Flights 62
GA #3: The Batplane 63
GUSTO 2: Flying Saucer Grows Wings 64
Plastics 69

Chapter 5 High Speed 71

Rex and the CL-325 71
SUNTAN 74
More Studies 77
GUSTO Supersonic Designs 78
Blip-Scan Study 82

Chapter 6 Competition from Convair 85

Super Hustler 86
FISH: First Invisible Super Hustler 90

Chapter 7 Archangel I 95

The Iron Maiden 100
Funding for GUSTO 100

Chapter 8 New Ideas 103

CHAMPION: Navy Inflatable Designs 103
Archangel II 106
Weekend Work: A Small Airplane 108
Land Panel 112
The A-3 113
Cherub 115
Final Design 116
Weight Reduction 118

Chapter 9 November 1958 Land Panel Review 121

Choosing FISH 121
More Funding 122

TABLE OF CONTENTS ix

White House Approval 123
Second Thoughts 123
Requirements 124
Comparison. 124
Further Studies 125
Backchannel 127

CHAPTER 10   LOCKHEED STEALTHY DESIGNS 129

The A-4 129
The A-5 133
The A-6 134
Arrow: Lockheed's FISH 137
Nonstealthy Designs 138
The A-7: Ugliest Airplanes in the World 139
Supersonic Refueling 140
Eisenhower Meets the Flying Saucer 141

CHAPTER 11   REFINING FISH. 143

Facilities. 143
Subsystems 147
New FISH 148

CHAPTER 12   LOCKHEED'S LARGE AIRPLANES 155

The A-10 155
The A-11 155

CHAPTER 13   JUNE 1959 LAND PANEL REVIEW 159

Demise of FISH. 161
Changing Directions 162
KINGFISH 164
The A-12. 172
Minority Report 175
Resurrecting FISH. 176
More Agency Studies 176
Framing the Decision 178
Informing the President 180
Final Decision 181
Ramping Down Convair 183
OXCART. 185

Chapter 14 Proving the A-12 ... 187

Radar-Absorbent Materials ... 196
Ionized Exhaust ... 203
Progress ... 204
Full Go-Ahead ... 208
Design Revision ... 208
The PDP-3 ... 210
Late Changes ... 211
The Dish ... 213

Chapter 15 Propulsion ... 217

The J58 Engine ... 220
Designing for a New Mission ... 221
More Changes ... 223
Simulations ... 226
Flight Test ... 228

Chapter 16 New Countermeasures, New Threats ... 231

Project KEMPSTER ... 231
TALL KING ... 234
First Mission ... 235

Chapter 17 Conclusion ... 237

F-12 ... 237
SR-71 ... 240
M-21 ... 241
A-12 Bomber Variants ... 242
Convair WO 540 ... 243
Lockheed D-33 ... 243
Project ISINGLASS ... 244
B-2 Stealth Bomber ... 244
Summary ... 244

Appendix A Skunk Works Engineering Staff ... 247

Appendix B Timeline of Projects RAINBOW and GUSTO ... 249

Appendix C Supporting Materials ... 261

GUSTO Documents Archive ... 261

REFERENCES **263**

BIBLIOGRAPHY **271**

Notes on Sources *271*
Drawings *271*
Audio Recordings *272*
Articles *272*
Reports and Presentations *273*
Patents *274*
Books *274*
Interviews by Author *274*
Telephone Interviews by Author *275*
Electronic Mail to Author *275*

INDEX **277**

PREFACE

I first became interested in the genesis of the Lockheed A-12 Blackbird more than 10 years ago. It appeared to have been one of the most difficult engineering projects ever attempted. As an engineer who has seen projects drag out as customer requirements changed, I wondered what had happened to require the Skunk Works to work through 12 different designs before the customer was satisfied with the product.

As I began to interview participants and their next of kin, I realized that there were more dimensions to the story, and its starting point was pushed back earlier in time. First, I was urged to tell the stories of the many Lockheed employees who had labored in the shadow of Clarence L. "Kelly" Johnson. Next, I realized that the designs were driven by the competition between Lockheed and Convair under Project GUSTO. Finally, I was able to get the stories of the scientists and engineers from Massachusetts Institute of Technology Lincoln Laboratory and the Scientific Engineering Institute (SEI) who initially worked under Project RAINBOW to make the U-2 invisible to radar and then influenced the design of the follow-on aircraft.

The result has been a history of the origins of stealth technology and how it was applied in the design of the world's fastest jet. It attempts to show when and where various concepts were developed, which ideas were rejected and which were used, and how the various participants interacted.

As with any history, the story presented here is influenced by the material available, whether it is the selectively released documentation from the Central Intelligence Agency or the sometimes contradictory remembrances of the participants. I can only hope that this narrative is close to the truth. As more documentation comes to light, the story might continue to evolve.

Paul A. Suhler
Irvine, California

Acknowledgments

I am indebted to a great many people for their help with this book. First I must acknowledge Jay Miller for his encouragement, advice, and friendship through the years, as well as enabling my access to the Jay Miller Collection, which is maintained in Little Rock, Arkansas, by the Aerospace Education Center. The Aerospace Education Center deserves a great deal of credit for its work in conserving the collection while working on a limited budget.

Robert H. Widmer, who as the head of advanced development at Convair was Kelly Johnson's counterpart during the design competition, was generous with his time and provided me with his unique viewpoint on the work. Robert Baldwin, the son of Lockheed's Edward Baldwin, allowed me access to his father's papers and was the first to urge me to tell the story of the little-known individual engineers. Chris Pocock, the historian of the U-2, provided me with advice, technical information, and introductions to numerous people. Peter Merlin, of NASA, has provided information and advice, all of the way back to that first hike through the Nevada desert to an A-12 crash site.

The late Herbert Rodgers, brother of the late Frank Rodgers, supplied Frank's memoir and introduced me to Frank's colleagues, thus opening up a major part of the story. I was gratified that Herb was able to read a nearly final version of this book and to see that his brother's contributions were recognized.

Many of the original participants were kind enough to allow me to interview them in person, by phone, and often by e-mail: General Leo P. Geary, F. Robert Naka, Thomas C. Bazemore, Norman H. Taylor, Richard Leghorn, Daniel Schwarzkopf, Brint Ferguson, and Robert Butman. Many former members of the Lockheed Skunk Works spoke with me, including Edward Lovick, Henry Combs, Elmer Gath, Norman Nelson, Bill Bissell, Frank Bullock, Charles VanDerZee, Bob "Shop" Murphy, Bob "Flutter" Murphy, Herb Ermer, Frederick Schenk, William Taylor, Sam Kelder, Sherre Lovick, and Alan Brown. Many had distinguished careers that would be worthy of their own books.

Several members of the Pratt and Whitney performance group gave me their recollections of the development of the model 304 and J58 engines: Norman Cotter, Robert Abernethy, Tom Warwick, and Bob Boyd. Other participants in projects AQUATONE and OXCART who have helped include Albert D. "Bud" Wheelon [former deputy director of the Central Intelligence Agency (CIA) for science and technology]; pilots James D. Eastham, Mele Vojvodich, Frank Murray, and Marty Knutson, and EG&G engineers Wayne E. Pendleton, and T. D. Barnes.

Others who have provided me with information include Jeanette Remak, James Goodall, Jeffrey Richelson, Jeffery Hartley and the staff of Archives II of the National Archives and Records Administration, David Haight of the Dwight Eisenhower Presidential Library, John Wilson of the Lyndon B. Johnson Presidential Library, Gordon Bell, John Stone, James Gibbs, Eric Nystrom, David Rodgers, David Lednicer, Chad Slattery, Allyson Vought, Randy Kent, and Donald Welzenbach (co-author of the CIA's history of overhead reconnaissance). I am grateful to Joseph Donoghue for introducing me to the National Archives, for sharing the fruits of his own research, and for sponsoring me for membership in Roadrunners Internationale.

The late Roger Cripliver is worthy of note as the unofficial historian of Convair. At his own expense, Roger maintained an enormous archive of documents that would otherwise have been discarded. Although he never achieved his initial goal of writing a book, he took great satisfaction in having provided information to the authors of over 20 books on aviation history.

Finally, I'm grateful for the help of my wife, Linda Maher. Through the years she has encouraged and helped me with this work, whether it was by transcribing interviews, trekking through the desert, or giving advice based upon her own experience as an author of legal works, screenplays, novels, and short stories.

Chapter 1

INTRODUCTION

On a warm August day in 1956, four men sat in a convertible in a Boston parking lot with the top up and the windows closed. In the back were Franklin Rodgers, Robert Naka, and Thomas Bazemore, scientists in the Radar Division of Massachusetts Institute of Technology (MIT) Lincoln Laboratory. In the front seat was Edwin Land, known to them as the founder of the Polaroid Corporation and an advisor to the government on various defense projects. Until this moment, however, they had never heard of one particular highly secret project.

> For the past six weeks the U.S. has been flying a reconnaissance aircraft over Russia at altitudes far above the reach of their air defenses. The good news is that we have been able to go anywhere we wish without fear of being shot down. The bad news is that their radars have been able to track us continuously from border to border. As a result, they have been able to scramble their interceptors against us but even though they have climbed to their maximum altitude or gone vertical until they stall out just before launching their air-to-air missiles, they have not been able to come within several thousand vertical feet of our aircraft. Since they know we're there, it can only be a matter of time before they improve the reach of their interceptors and/or missiles making it possible for them to knock us out of the sky. Our only hope would appear to be to make our aircraft invisible to their air defense radars (Rodgers, Franklin A., unpublished memoir, 3 Feb. 1995).

At the time, Land did not mention the Central Intelligence Agency (CIA), the Air Force, Lockheed Aircraft Corporation, or Pratt and Whitney Corporation, but it was the beginning of an audacious effort that would involve all of them and thousands of people more. The end result would be the fastest and highest flying jet the world has seen.

In 1956, the shock of the 1941 Japanese attack on Pearl Harbor and the horrors of the war that followed were still fresh in the minds of America's leaders. When the Soviet Union exploded its own atomic bomb in September

1949, the sense of vulnerability increased, with the realization that the next surprise attack could destroy American cities and kill millions of people. Deterring an attack required a credible defense. But that required knowing not just when the attack would come, but what form it would take. President Dwight Eisenhower needed hard numbers on Soviet bomber strength, and reliable numbers were not to be had.

When Eisenhower turned to the Department of Defense (DoD), he was told that there was a "bomber gap" in which the United States lagged behind the Soviet Union. DoD was concerned with defense and did not want to be blamed for underestimating the other side's capabilities. Although the defense industry believed in the need for the bombers and other weapons that they made, they also needed business and supported the DoD's position by lobbying Congress. Eisenhower, on the other hand, had broader responsibilities than industrial profit or jobs in a few states. He understood how the military-industrial complex would always ask for more money than the country could afford. The Central Intelligence Agency was formally charged with learning the Soviet Union's capabilities and intentions, but the Union of Soviet Socialist Republics (USSR) was a very difficult target to penetrate using traditional espionage techniques.

The other players in the intelligence game were scientists from industry and academia. The Office of Defense Mobilization had asked the president of MIT, James R. Killian, Jr., to form a group that became known as the Technological Capabilities Panel (TCP). The TCP consisted of three "projects": the first focused on U.S. offensive capabilities, the second on defensive capabilities, and the third on intelligence capabilities.

Project 3 was led by Edwin H. "Din" Land, the founder of the Polaroid Corporation. Following Land's "taxicab rule"—that to be effective, a working group had to be small enough to fit in a taxi—there were only five members. Astronomer Jim Baker and physicist Ed Purcell were from Harvard University, chemist Joseph Kennedy was from Washington University, and mathematician John Tukey was from Bell Laboratory and Princeton University. Their job was not to provide intelligence per se, but to help find technical means to obtain that intelligence. One of the technical means they decided on was aerial reconnaissance.

"A UNIQUE OPPORTUNITY"

Richard Leghorn had been involved in aerial reconnaissance for almost 20 years. Upon graduating from MIT in 1939, he had been commissioned in the Ordnance Corps but had decided that he would actually serve in reconnaissance. He was able to transfer to the Army Air Corps and not only worked in the Aeronautical Photographic Laboratory at Wright Field, but also flew photo reconnaissance missions in combat. After the war he argued for pre-D-Day photography, most famously at the 1946 dedication of the Boston University Optical Research Laboratory (BUORL).

His thesis was that if more than one country, that is, not only the United States, had nuclear weapons, then it was essential to have prior warning of the possibility of an attack. Once a nuclear attack had been launched, it would be too late. Aerial reconnaissance could look for signs of potential hostilities, such as mining of radioactive ores and building production facilities for fissionable materials.

Leghorn understood the political constraints against such reconnaissance. He conceded that overflight without permission of the target country was a violation of international treaties and would be unlikely to be permitted. Nevertheless, he proceeded to outline the technical means, which he saw as an extremely high-altitude aircraft camouflaged against visual observation. He also noted that "It is not inconceivable to think that means of preventing telltale reflections of other electromagnetic wave lengths, particularly of radar frequency, can be developed" [1].

In 1952, Leghorn had participated in the Beacon Hill Study, which analyzed the requirements for reconnaissance prior to the commencement of hostilities. It identified sensors and aircraft technologies to carry them, including improved cameras and balloons and airplanes capable of flight at over 70,000 ft. He refined these ideas further in early 1953, while working for Colonel Bernard "Bennie" Schriever in the Air Force's Development and Advanced Planning (AFDAP) office at the Pentagon, where he produced an Intelligence and Reconnaissance Development Planning Objective. Leghorn then left active duty and returned to private industry. He would not reenter the reconnaissance world for several years [1].

Historian Chris Pocock has documented the twists and turns that Leghorn's ideas took through the government bureaucracy [2]. By March 1954, the Air Force had awarded a contract to Bell Aircraft to develop a twin-engine reconnaissance aircraft and another to Martin to adapt the B-57 bomber for high-altitude operation. It had also rejected Lockheed's CL-282 single-engine design (Fig. 1), mainly because it did not fit the Air Force's preconceived notions of what a military airplane should be.

Less than two months later, the Soviet Union flew a long-range jet bomber at its May Day celebrations, adding to the existing anxiety over the Soviet's missile development program at Kapustin Yar.

Lockheed's concept of a lightweight single-purpose aircraft was rescued by Assistant Secretary of the Air Force Trevor Gardner and by Marine Corps Brigadier General Philip Strong, who was serving as the chief of operations of the CIA's Office for Scientific Intelligence (OSI). Strong took the concept to Land's TCP Project 3.

With advice from Cornell University aerodynamicist Allen Donovan, the members of Project 3 evaluated the CL-282, and on 5 November 1954, Land

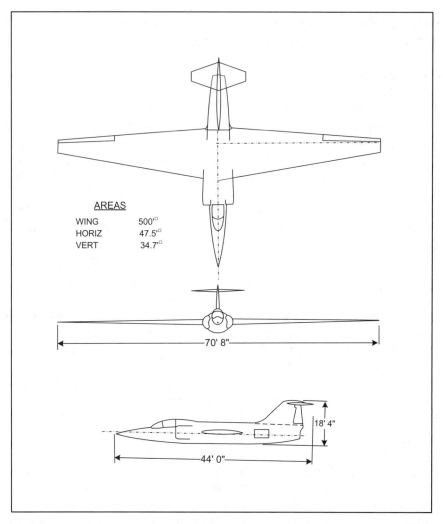

Fig. 1 Based upon the F-104, the CL-282 was Lockheed's first proposal for a high-altitude reconnaissance aircraft. (Drawn by author from Lockheed Aircraft Corporation drawing.)

wrote on their behalf to Allen Dulles, the Director of Central Intelligence. His memorandum, "A Unique Opportunity for Comprehensive Intelligence," said that while aerial photography

> ... could be the most powerful single tool for acquiring information, it has until now been dangerous to fly over Russia. Up till now, the planes might rather readily be detected, less readily attacked, and possibly even destroyed. Thus no statesman could have run the risk of provocation toward war that an intensive program of overflights might produce [3].

They recommended that the "... CIA, as a civilian organization, undertake (with the Air Force assistance) a covert program of selected flights. Fortunately a jet powered glider has been carefully studied by Lockheed Aircraft Corporation for overflight purposes" [3].

Land laid out how the Agency (the CIA) should organize its task force, the details of the aircraft, the targets it should photograph, and the schedule and cost. The memo also included two concepts that within two years were to prove incorrect. The plane was "... so obviously unarmed and devoid of military usefulness, that it would minimize affront to the Russians even if through some remote mischance it were detected and identified" [3].

Dulles, who had held earlier discussions with Eisenhower on aerial reconnaissance, made his recommendation in a memo to the President that the RB-57 be developed as an interim measure, that a "specially designed reconnaissance aircraft with more advanced performance characteristics" be simultaneously developed for operations at 70,000 ft, and that a night reconnaissance program be conducted at low altitudes [4]. On 24 November, Dulles and five others met with the President to received his formal approval.

Dulles assigned the project to his Special Assistant for Planning and Coordination, Richard M. Bissell, Jr. A brilliant Yale-trained economist, Bissell had shown his capabilities as the deputy director of the Marshall Plan and had run its day-to-day activities.

As would be the norm for Bissell's projects, work began immediately with a verbal approval, and the formal documents followed later. It was 7 January 1955 before the project plan was sent to Dulles for a written approval. The overflight project as a whole was assigned the cryptonym AQUATONE, and the development of the aircraft by Lockheed was a subproject named OARFISH. Five other subprojects included photoreconnaissance equipment, electronic equipment, photo intelligence analysis, electronic intelligence analysis, and pilot recruitment and training.

Lockheed assumed responsibilities that for any other aircraft would have belonged to the customer. The plan stated that Lockheed was "... willing to take full responsibility for the design, mock-up, building, secret testing, and field maintenance of this unorthodox vehicle. It therefore appears entirely feasible for a CIA task force to undertake a covert overflight program based upon the CL-282 ..." [5]. It was 25 February before Dulles received definitive contract SP-1913 for Lockheed to produce 20 aircraft [6].

DESIGNING THE U-2

The son of Swedish immigrants, Clarence L. "Kelly" Johnson graduated from the University of Michigan in 1933 with a master's degree in aeronautical engineering. He landed a job with Lockheed Aircraft Corporation in Burbank, California. Initially working for Hall Hibbard, he worked on most of Lockheed's most famous aircraft, such as the Electra, the P-38, and the Constellation. During World War II, he pulled together and led the small group that designed the XP-80, which became the first production jet fighter in the United States. During that project, the group adopted the name "Skunk Works," which was taken from the Al Capp comic strip "Lil Abner" (see Fig. 2). After the war the group worked on other aircraft, including the F-104, but not again in the same mode of secrecy until the U-2 began.

Johnson's right-hand man was Dick Boehme, a highly competent engineer whose calm temperament contrasted with Johnson's sometimes fiery one. Ed Baldwin, one of the chief designers, remembered that on the Tuesday before Thanksgiving of 1954,

> Dick came to me and said, "I'm going to ask you a question, then I want you to forget that I did." He asked who we had that I would recommend for a Skunk Works operation, and I said, "Me." I couldn't let an opportunity like that go by. I recommended Ray Kirkham and a couple of others. He said, "OK, thanks. Now forget that we ever talked," so I knew something was up.

Three days later, Russ Daniel told Baldwin and four others to be in Kelly's office at 1300 hrs. The others were Carl Allmon, Elmer Gath, Bob Wiele, and

Fig. 2 The original stuffed skunk from the Lockheed Skunk Works. (Courtesy of the family of Edward P. Baldwin.)

Henry Combs. They were the first to be briefed on the new project. By the following Monday, 22 engineers had been recruited and eventually there were 28 (Baldwin, Ed, "U-2 General Arrangement," unpublished memo, undated). Bill Bissell recalled that when Johnson briefed his group, "He said, 'Eisenhower wants a new airplane to take photos of Russia.' Everyone's mouths fell open" (Telephone interview with Bill Bissell, 13 June 2005).

Others were not part of the project. Bill Taylor was working on the F-104 and knew that Baldwin and others had gone to work on a new project that they were not talking about. Eventually assembly of the U-2 prototype began before all of the security partitions were up in the shop, and Taylor caught sight of a wing, and so he at least knew that they were building an airplane (Interview with William Taylor, Hollywood, CA, 17 Nov. 2003).

One of the ways the Skunk Works produced designs quickly was by not redesigning things that already worked. In the days before computer-aided design systems, lofting was a tedious manual process of producing drawings of every fuselage structural component. The drawing for each component would be scribed on a metal panel, which would be inked and used to produce blueprints for production. These panels were called lofting boards.

Because the CL-282 had been based on the F-104, using its lofting boards was an obvious way to cut at least some of the work. However, the F-104 lofting boards were in use, and they could not be borrowed without raising questions. The XF-104 lofting boards, on the other hand, were not in use and thus became the basis for the design of the U-2.

While Gath pulled together information on the J57 engine, Baldwin began the three-view drawing (see Fig. 3). Also known as a general arrangement (GA)

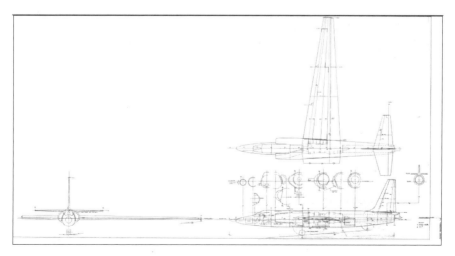

Fig. 3 The U-2 general arrangement (GA) drawing was completed by Ed Baldwin on 29 January 1955. (Courtesy of the family of Edward P. Baldwin.)

drawing, the three-view was the first step in the design of a new aircraft. Using the dimensions for the XF-104, he drew the lines back to the rear pressure bulkhead. Johnson did not want the XF-104's pointed nose, and so Baldwin used a French curve to put on a round "Lockheed-type" nose. Gath

> ... gave me a 1/20 scale drawing of the J57 engine so I could locate it in the area of the wing carry-through structure close to the center of gravity. I also knew that the depth of the frame at the side is a function of the wing attachment, this even before we even had any basic loads. I knew how far apart the bolts were for taking the bending versus the depth of the ring through that area, and using that same proportion, I determined the clearance around the engine. Elmer told me how much room he needed to roll the engine in, so I left clearance on either side for that, and struck a circular fuselage with what would be a wing fairing, and then cut it off straight.
>
> The 1/20th scale General Arrangement drawing was the first drawing done for design. It is drawn to actual scale, and it is the drawing from which all the other drawings came. Carl Allmon used it to do the lofting, representing the curves on the drawing with mathematical formulae. We also used it to make a 1/10 scale 'bones' drawing, to actual scale as near as we could tell with the loads we knew at the time, and everyone worked from this to make the drawings for the bulkheads, the cockpit, and so forth. (Baldwin, Ed, "U-2 General Arrangement," unpublished memo, undated.)

The U-2 made its first flight on 4 August 1955 at a primitive test site about 85 miles northwest of Las Vegas, Nevada. To provide a legal basis for keeping out the uninitiated, the site had been designated as a part of the Nuclear Testing Site. It used the radio call sign "Watertown Strip," but was more often known as "the Ranch," a shortening of Kelly Johnson's euphemistic name "Paradise Ranch."

By 1956 the U-2 was ready for operational missions, and a number of important relationships had been established. Within the CIA, Richard Bissell's deputy was an Air Force colonel with a degree from CalTech, Jack A. Gibbs. Gibbs's experience as an aviator was essential in organizing the U-2 operational detachments, and his engineering acumen was important in the work leading to the follow-on aircraft. Another principal in Bissell's office was James A. Cunningham, the chief of the administrative branch. A graduate of Brown University, Cunningham impressed others as a "mental packrat." He often found himself indirectly refereeing between Bissell and Johnson when they were at odds.

Most of their day-to-day interface with the Air Force was via Lt. Col. Leo P. Geary. Geary had been associated with the Agency for a number of years;

from 1951 to 1954 he had run an air section for the CIA in Greece. In June 1955 he began supporting the CIA from within the Pentagon and had to maintain a low profile and keep the U-2 program in business without becoming embroiled in politics. As a lieutenant colonel (and later a colonel), he did not have enough power to take on generals who might take offense at the involvement of the CIA in aviation. In the process of working on the U-2, he got to know Kelly Johnson, and over the years the two became "… the closest of friends" (Telephone interview with Leo P. Geary, 31 July 2002).

Geary also formed a great deal of respect for Bissell and was extremely impressed with his intellect and his ability to learn a new subject seemingly overnight. The respect was reciprocated; when Geary once ordered some U-2 upgrades in Bissell's absence, Bissell was initially angry, but recovered once he understood that it was the right decision. Forty years later, Geary was still angry over Bissell's being forced to resign because of the failure of the Bay of Pigs invasion.

A major reason for the success of the U-2 and follow-on programs was because of the mutual trust among Bissell, Geary, and Johnson.

Chapter 2

RAINBOW

From the beginning of Project AQUATONE, Eisenhower had been very concerned that the U-2 should not be detected by Soviet radars. He knew that overflights would be considered an act of aggression and might even lead to war. The first overflight of denied territory took place on 20 June 1956 and photographed East Germany, Czechoslovakia, and Poland. Over the next week the Agency (CIA) evaluated not just the photos returned but also communications intercepts and the S-band recordings made by the U-2's System One electronics intelligence (ELINT) receiver in an attempt to determine whether the flight had been tracked. Signals from numerous Token radars and some from an unknown type, one having a conical scan, were recorded. There were no indications that any one unit had followed the U-2, but "... the intensity of the signals from Tokens near the path of flight is such that it is believed that echoes must have been received on the scope of the Token [7].

The U-2's altitude was apparently misread as 42,000 ft.

As the overflight program moved forward, Eisenhower's anxiety over detection increased. The second and third missions over the Eastern bloc launched on 2 July, and Eisenhower told his aide, Colonel Andrew Goodpaster, that it was urgent that he be informed whether the aircraft had been picked up. On 3 July, the day before the first flight over the USSR, Eisenhower had Goodpaster tell Bissell that 10 days' worth of missions were approved, but that he was to provide interim reports on tracking and attempted interception [8]. And after the successful completion of that first mission, he had Goodpaster tell Director of Central Intelligence (DCI) Allen Dulles that "... if we obtain any information or warning that any of the flights has been discovered or tracked, the operation should be suspended until he (the President) has an opportunity to consult on the matter with Secretary [of State John Foster] Dulles and Allen Dulles" [9].

After being told this, Allen Dulles and Bissell went to the White House to clarify the message. They told Goodpaster that it would be at least 36 hours after each mission before they would receive the first reports of detection, tracking, and attempted interception, and that it might be as long as several days or weeks. If they had to wait this long between each mission, they would not be able to follow the President's intention of covering as many targets as quickly as possible. Goodpaster told them that he understood that the President wanted the missions to "... go forward at the maximum rate until the first evidence of tracking was received ..." [10]. On 10 July, Goodpaster confirmed this with the President and informed him of some indications of tracking. Goodpaster told this to Dulles, along with his impression that the President seemed very close to stopping the overflights [11].

The same day, the Soviets delivered a protest note, and Eisenhower ordered a halt.

In the week after the shutdown, Herbert Miller wrote a memo to Bissell summarizing the achievements of the AQUATONE program to date. This was probably intended to give Bissell ammunition in arguing for a resumption of flights, or at least to defend it from a complete termination. Miller's thesis was that the benefits far outweighed the potential damage to international relations.

He began by differentiating the U-2 from a military reconnaissance aircraft, saying that it was merely a substitution of an airplane for a traditional agent and that as such it was not a target-spotting function like one the military would carry out as a prelude to an attack on the Soviet Union. He emphasized that the intelligence gained covered many aspects of Soviet culture, not merely the location of targets.

Miller then reviewed particular things learned, beginning with the fact that large fleets of bombers had not been found at the five air fields covered, even though "regiments" of bombers had been expected at two of them. Coverage of naval facilities at Leningrad showed new submarine ways at one yard and the fitting of an additional Sverdlov class cruiser. The missions also showed an army installation at Bykhov and nearby training grounds.

But most important, the flights covered 400,000 square miles broadly and up to 15,000 square miles in high detail, providing hard evidence of the state of the economy and military of the USSR. This showed previously unknown features, such as new army bases, airfields, and large industrial installations. And he pointed out that that the area beyond the Ural mountains had yet to be surveyed.

He concluded:

> Five operational missions have already proven that many of our guesses on important subjects can be seriously wrong, that the estimates which form the basis for national policy can be projections from wrong guesses and that, as a consequence, our policy can indeed be bankrupt. In this light, the danger to our international relations appears to be far greater if we do not carry out the AQUATONE plan than if we do carry it out, having laid sufficient ground work to assure that the interpretation of the activities is as an intelligence operation rather than as a reconnaissance prelude to hostilities [12].

BEGINNING THE SEARCH FOR INVISIBILITY

Eisenhower was discouraged by the ability of the Soviets to track the U-2. On 19 July, Allen Dulles saw him and reiterated that no new missions would be flown without consulting the president. Eisenhower pointed out that he had been told that only a very small percentage of the AQUATONE and peripheral flight missions would be detected, which had not proven to be correct. He said that America's reaction to overflights by the Soviets would be "drastic" and that, although he was concerned by protest notes, he was much more worried about a loss of confidence by the American people [13].

On Thursday, 16 August, Bissell convened a meeting to find a way to solve the tracking problem. Kelly Johnson later wrote, "Up to see Land, Purcell, Stew Miller with Herb & Dick. Worked till 1:30 and two bottles of Scotch. Up at 7:00 and we resumed—even Dick. By noon we had program 'X' going. My biggest job now" [14].

Land took on the job of recruiting a team of experts in radar to work on the problem. The obvious place to go was Lincoln Laboratory, a part of the Massachusetts Institute of Technology (MIT). Lincoln Lab was the successor to the MIT Radiation Laboratory, or Rad Lab, the center of American research in radar during World War II. In 1950, the Project Charles summer study at MIT had recommended the establishment of a national laboratory to work on air defense; Lincoln Lab was founded and did work for the Air Force, Army, and Navy. Two of the Lab's major achievements were the Semi-Automatic Air Ground Environment (SAGE—an integrated system of radars and anti-aircraft weapons linked to computers) and the Distant Early Warning (DEW) Line, which scanned northwards for Soviet attacks over the arctic. Lincoln Lab was not merely a think tank; it did the basic science for air defense concepts, developed practical systems, and oversaw their production and installation.

Land met with Marshall Holloway, the lab director, who summoned Frank Rodgers, associate head of the Radar Division, and Bob Naka and Tom Bazemore, members of groups specializing in radar transmitters and receivers. At 1000 hrs, the three found themselves waiting outside Holloway's office, wondering why they were there. Holloway called them in and introduced them to Land. Rodgers had worked with Land's company, Polaroid, some months earlier about possibly using their instant film for a radar display application, and his first thought was that Land was back to talk about it some more.

Without saying what was going on, Land led the group out of the building, so it was obvious that something else was up. They drove away from Lincoln Lab, to an empty parking lot at the Cambridge Research Center, an Air Force laboratory that was still under construction. Land told them what they needed to know—that they had to solve a radar problem—and did not tell them what they did not need to know—who all of the players were. That knowledge would come later, in bits and pieces.

Land asked whether they thought that an airplane could be made undetectable. They were skeptical but agreed to try. Land gave them the usual instructions about the sensitivity of the project and that they could not mention it to their wives or even to Marshall Holloway. Although Lincoln would continue to pay them, they would be working on this new project full time, and somehow Lincoln would be reimbursed. That compartmentalization away from the leaders of Lincoln Lab would eventually affect the evolution of the project.

Within a few days, Rodgers was contacted by Herb Miller and summoned to a meeting in Land's office in Polaroid's building in Cambridge. Unlike the steel-and-glass technology parks of 40 years later, the building was red brick and dated from the 19th century. Land's door was unmarked and unpretentious.

Rodgers was surprised to be introduced to one of the people present. He had known Ed Purcell only by reputation, as a Nobel laureate and as the editor of a volume in the Radiation Laboratory Series, a collection of technical articles that was the bible for radar engineers. Rodgers did not know the man who looked a bit like Abe Lincoln with glasses, who constantly fidgeted with a paper clip, and who was only introduced as "Mr. B." Land turned his telephone dial and jammed it with a pencil, explaining that it was a precaution against one type of telephone bug.

The meeting turned out to be a brainstorming session. Purcell proposed applying a material to the outside of the airplane which would consist of a printed circuit pattern that would diffuse the energy of a radar beam illuminating the airplane. They talked it over and figured out a quick way to fabricate a sample for a test. Rodgers was to build and test the sample.

At Lincoln, Rodgers, Naka, and Bazemore began to pull together the people and equipment to do the job. At the same time they had to keep it secret from virtually all of the staff, who had no need to know. Forty years later Frank Rodgers remembered:

> There happened to be a small, wooden shack on the roof of the building in which my radar groups were housed. It probably had been a contractor shack during the construction of the building which had never been removed. It showed no evidence of having been occupied since the construction was completed so I was able to move in with no one the wiser.
>
> The equipment requirements to get started were quite modest and I was able to "steal" what I needed from various groups which reported to me. Staff was more of a problem. It was bad enough that the associate head of a major division was dropping out of sight. If some of the best talent reporting to him also started to disappear, it could only raise more eye brows. I succeeded in recruiting one of the best experimentalists in my division. Danny Schwarzkopf was an absolute whiz whether working on radar transmitters or receivers. But another outstanding man I badly wanted on my team, Dr. Ed Rawson, was in another division. It would not be easy to shake him loose.

I decided to try a frontal attack. I went to the head of that division and asked very politely if I could "borrow" Ed for awhile. He asked, "For what?" I committed hari kari [*sic*] by responding, "I'm sorry but I am not permitted to say." The man went into orbit. I finally asked him to check with Holloway on whether my request was appropriate or not. Within a day, Rawson was on my team but his former division head had me on his "most wanted" list from then on. (Rodgers, Franklin A., unpublished memoir, 3 Feb. 1995.)

The "shack" on the roof of B Building is shown in Fig. 4. The window was covered and models and test materials suspended in front of the window for illumination by radar equipment. At the left of the picture, on the roof of C Building, was a radar dish covered with one of the first geodesic domes.

Before Schwarzkopf and Rawson, Rodgers recruited an antenna theorist. He is a bit of an enigma because he will not be interviewed and will not allow his former colleagues to identify him. His is a key part of the story because he contributed a solution to an important problem for both the U-2 and the follow-on aircraft.

Rodgers was also able to recruit Bob Butman, the leader of another group in the Radar Division.

Fig. 4 MIT Lincoln Laboratory in the mid-1950s. The arrow indicates the "shack" on the roof of B Building where the Project RAINBOW anti-radar work began. (Reprinted with permission of MIT Lincoln Laboratory, Lexington, Massachusetts.)

One of the difficulties the project encountered was that its secrecy was contrary to the way Lincoln Lab was organized. Every Lab employee had a secret clearance, and projects were not compartmentalized. This meant that everyone could discuss his work with any other employee, which meant that a casual conversation with a person from another project could lead to an unexpected solution to a problem. Because RAINBOW was top secret and compartmentalized, the team members could talk with no one outside the team, and even the steering committee (of which Rodgers was a member) could not hear about the work. This placed stress on many of the members' working relationships.

STEALTH IN 1956

The team members were not beginning from absolute zero in their work. There had already been a decade and a half of theoretical and experimental work on reducing the radar cross section of vehicles. At the Rad Lab during World War II, Winfield Salisbury had developed a coating designed to "... prevent or reduce reflection of electromagnetic radiation from surfaces." In one application the "Salisbury screen" (Fig. 5), as it came to be known, consisted of the metal surface to be protected, a layer of wood, and a layer of graphite-coated canvas. The thickness of the wood layer (which would not interfere with the electromagnetic energy) was $\frac{1}{4}$ of the wavelength of the expected radar wave. The coated canvas would reflect a wave that would be

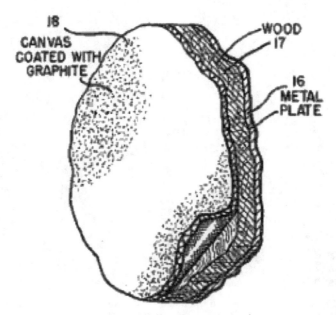

Fig. 5 The Salisbury Screen protected a metal plate with a $\frac{1}{4}$-wavelength wooden spacer and a layer of graphite-coated canvas. (Courtesy of U.S. Patent and Trademark Office.)

180 deg out of phase with the wave reflected by the metal, and the two waves would interfere and return very little energy to the transmitter. The problem is that it would work best for only one radar frequency and then when the wave was aimed directly at the surface.

Salisbury's patent application described a number of possible applications for the screen, including, "... to protect an airplane or ship or other object against radio echo detection systems of the enemy by treating the surface thereof or large portion of it so as to prevent reflection of electromagnetic waves of the particular wavelength used" [15]. Purcell's coating for the U-2 was a Salisbury screen. The outer layer of the coating would create a reflection that would interfere with the reflection from the metal skin beneath the coating.

In Germany, Johannes Jaumann had developed a compound of synthetic rubber impregnated with 10–20 µ-diam particles of iron oxide, which was applied to the snorkels and periscopes of U-boats, in an attempt to prevent their being detected by Allied aircraft and vessels protecting shipping convoys. The project was known as "Schornsteinfeger," or chimney sweep, a reference to the carbon black also used in the compound. There are claims that by the fall of 1944 carbon black was 90% effective in reducing the radar signature, in tests performed by the German Aeronautic Research Institute (DVL) using a 1.5-m airborne radar set. Jaumann went on to develop a layered radar absorber, which, like the Salisbury Screen, has found more applications and can be found today in catalogs of electrical supplies.

Also in Germany, the Horten Ho229 flying wing has been said to have stealth properties. Its wings were constructed of two thin layers of plywood glued together with a charcoal and sawdust mixture [16]. Although stealth was not a requirement of the 3X1000 project under which the development was funded, Reimar Horten apparently claimed after the war that the mixture was in fact added with the intention of reducing the aircraft's radar cross section (RCS), the measure of how much an object reflects radar, expressed as an area.

Beginning in 1950, Kip Siegel and other members of the University of Michigan Radiation Laboratory had performed a systematic study intended to understand the physics of radar reflection. They made RCS measurements of various geometric shapes and derived equations to describe the reflections. Eventually, the work would be funded by the Air Force and one of many published papers was "A Theoretical Method for the Calculation of Radar Cross Sections of Aircraft and Missiles" by Crispin et al. [17]. How effective the method was is not known. The first successful design that used accurate calculation of RCS during the design process was Lockheed's HAVE BLUE; it took another 15 years after the paper by Crispin et al. until computer processing power and numerical methods had advanced enough to be useful.

Others had observed that objects surrounded by a plasma produced a radar return much smaller than expected. In the same month that the antiradar work began at Lincoln, Arnold Eldredge, a scientist at General Electric in California, filed a patent application for a means to make an object like an aircraft invisible to radar. His technique was to put an electron accelerator in the aircraft and shoot a beam of particles to ionize the air molecules in front. The plasma cloud would flow around the aircraft and diffuse incoming radar energy. Years later, this scheme would be adapted to the follow-on to the U-2.

Mis-Proof of Concept

The Lincoln Lab team's work began not with an academic literature review, but with an experiment to test Ed Purcell's concept of treating the skin of the aircraft to absorb radar energy. In less than a week, they had fabricated the treatment and applied it to a small flat plate of aluminum. In classical controlled experiment form, they illuminated both a treated plate and an untreated plate with a radar unit and found that the treatment gave a reduction between 10 and 15 dB, which was quite significant. Rodgers reported back to Herb Miller, and "... the next thing I knew I was on an airplane headed to Burbank. My equipment was checked as baggage and the treated aluminum plate was being couriered separately by Miller who was a 'white knuckled' air passenger." (Rodgers, Franklin A., unpublished memoir, 3 Feb. 1995.)

Almost immediately the specter of a security breach appeared; when Miller and Rodgers arrived in Los Angeles, the checked bag with the radar equipment was not to be found. "There were a couple of anxious hours while Miller verbally abused every employee of the airline he could find. ... Finally, the baggage was found with no evidence that anything but innocent incompetence had been involved in its delay." (Rodgers, Franklin A., unpublished memoir, 3 Feb. 1995.) In a top secret project run by the CIA, paranoia was the order of the day.

Rodgers had never met Kelly Johnson before, but found that he

> ... looked, in fact, just as he has been described by others. He immediately reminded me of W. C. Fields with a slightly bulbous, pinkish nose—not quite as prominent as Fields. Today I would describe him as sort of a cross between Fields and Yeltsin. When he was talking about his projects he would project a happy countenance; when in deep thought he would scowl like Yeltsin. He was obviously very proud, as he had every right to be, of his many aerodynamic triumphs. He showed us through his Skunk Works where yet another U-2 was being assembled. (Rodgers, Franklin A., unpublished memoir, 3 Feb. 1995.)

Then Miller pulled out the treated plate, and Rodgers explained to Kelly the theory behind the concept. He assembled the equipment and repeated the experiment he had done in the Building B penthouse, showing that the treated plate gave a significantly lower return. Johnson immediately began to consider how it could be used in the U-2. He focused on the parameter that was most crucial, "the additional weight that the treatment would add to the U-2 but [he] was obviously willing to keep an open mind and pursue the possibilities the treatment offered." (Rodgers, Franklin A., unpublished memoir, 3 Feb. 1995.)

Rodgers and Miller left the treated plate behind for Johnson's engineers to analyze.

Back in Cambridge, Rodgers and Naka tried the experiment again but with a slightly different radar frequency, and it did not work at all. Once they had analyzed the problem, they found that the experiment was completely invalid. When the plate being measured was placed directly over the horn of the transmitter, a resonant chamber was formed, and the receiver measured the microwave oscillations inside the chamber rather than just the reflection from the plate. The treatment caused different oscillations than did the untreated plate; it was coincidental that the treated plate gave a much lower measurement. It had nothing to do with what a radar unit would see when illuminating an airplane in flight. In that case, slight changes in frequency would not have made dramatic differences in the returns.

They consulted Purcell and realized that the right way to do the experiment was to have the radar horn a relatively large distance from the plate. Dan Schwarzkopf was brought in to build some equipment for the revised test, although at that point he was not told what it was all about. Rodgers later remarked that it was fortunate that they got the 10–15 dB reduction; otherwise, the project would have died right there (Interview with Daniel Schwarzkopf, Stow, MA, 30 Nov. 2003).

U-2 Measurements

Treating the U-2 to avoid detection by the various Soviet radars first required the team to determine what its reflections looked like at various points in its flight path. Different frequencies of radar were used for early warning, tracking, and fire control. Each would illuminate the U-2 at different angles and distances. To get a large number of measurements from precise angles, models of the U-2 were measured at different frequencies and orientations to the radar beams.

Although materials and small models could be measured in the penthouse, RCS measurements were needed for actual U-2s to make sure that the models corresponded to reality. Some of these were to determine what the aircraft in flight looked like to a radar, and others were to determine how much of the return was caused by each component of the airplane.

To get started quickly, early flights were conducted against existing Air Force tracking stations using standard radars. Later, specially built radar systems that more closely simulated Soviet radars were used to get more precise data.

For the flight measurements, a Lockheed test pilot would fly an untreated U-2 from the Ranch to Boron, a tiny California town on the north edge of Edwards Air Force Base, where the Air Force maintained a radar installation. The U-2 would fly a prescribed course while the Lincoln Lab scientists operated the radar and recorded measurements at various ranges and angles. The number of data points were much smaller than for the models because the U-2 couldn't easily be turned at arbitrary angles to the radar. One maneuver pilots particularly disliked was yawing the aircraft 20 deg left and right of the flight path because the constant pressure on the rudder pedal was tiring.

Once the initial skin treatment for the U-2 was ready, it was placed on one side of the aircraft. During the flights over Boron, it would present first one side, then the other for comparison. At this time an SCR-584 radar set was in use. Because the plan position indicator (PPI) display could not show the magnitude of the returned signal, it was replaced with an oscilloscope. This presented its own problems because scintillation in the signal gave a noisy appearance, which made it difficult to determine the value accurately (Schwarzkopf, Daniel, e-mail to author, 22 Oct. 2007).

During one of the first trips to Boron to get measurements of an untreated U-2, Bob Naka and Tom Bazemore met with the colonel who was in charge of the radar site. The colonel told them that they were in charge. They shut down the plan position indicators and asked that the radar antenna rotation be stopped. The colonel balked at that, saying he couldn't surrender that authority to them. They called Herb Miller, whose job it was to get the program whatever it needed. He contacted a General Parker, head of the Air Defense Command, and the antenna was stopped. After that, there were no more arguments at Boron about authority.

Later, the same thing happened when Miller, Dan Schwarzkopf, and Ed Rawson visited another Air Force base. When they arrived, the base commander said that he hadn't gotten permission to hand over control of the radar unit. Miller pulled a bottle of Jack Daniels out of his briefcase and said, "Let's go to the cafeteria and I'll make a phone call and we'll all have a drink." Miller's routine was well established because only 10 minutes later a captain brought in a teletype message. The colonel read it and said, "Well, as near as I can tell from this, if you guys want to pack up the radar site and put it in a truck and take it away, you're welcome to it." (Interview with Daniel Schwarzkopf, Stow, Massachusetts, 30 Nov. 2003.)

Borrowing standard radar sets at existing installations was only good for a start. Before long the team had established a facility at a military base in Daggett, California, where they could set up a system with better instrumentation.

Rodgers wrote that,

> My small team of Lincoln scientists began to see evidence of the faith the Agency placed in our endeavors and the lengths to which it would go to

provide us with anything for which we expressed a need. Many of the Russian radars were modeled after lend-lease early-warning radars the U.S. had provided the Russians during World War Two. Our military had long since retired such radars in favor of "more sophisticated" designs. But the old radars were proving surprisingly useful to the Russians in their effort to track the U-2. We informed Miller that we needed a radar which would approximate the capabilities of the Russians in this area.

The Agency was unable to locate any such radar still in the U.S. military inventory so it contacted someone high in the RCA Corporation and asked for a TV transmitter which RCA at the time was manufacturing and which we could modify to satisfy our needs. As luck would have it, such a transmitter was, at that very moment, sitting on a loading dock waiting to be trucked to an RCA customer. The RCA official made a few phone calls. Within hours, an Agency truck backed up to the loading dock and whisked the transmitter away. We never did hear how RCA explained to its customer why it did not get timely delivery of its equipment. (Rodgers, Franklin A., unpublished memoir, 3 Feb. 1995.)

The television transmitter (channel 4) was hooked to an array of four Yagi antennas mounted on a telephone pole, to simulate a low-frequency search radar.

Ground and flight measurements were made at Indian Springs (now Creech) Air Force Base, northwest of Las Vegas. The arrangements for people and equipment had been made in a 26 October phone call from Jack Gibbs to a Lt. Col. Gordon at Air Force Headquarters. The equipment included an SCR-584 radar set, an SD-3 radar set, and an identification friend or foe (IFF) interrogation set. The people were six Air Force radar technicians to maintain the SCR-584. He also requested access to the base for project personnel. The work was to start on 15 November [18].

A pole that could lift and rotate an entire U-2 was installed. In this controlled situation, measurements could be made with much greater accuracy in the aircraft's orientation than could be done in flight. To eliminate the reflections from selected parts of the aircraft, horsehair batts were attached to absorb the energy.

The Lincoln Lab penthouse was not an ideal place to work, and not only because of its small size. On the roof of another building was a big 400-MHz radar dish, which was sometimes aimed at B Building to provide noise to test another group's computer that was being hardened against interference. When this happened, all of the equipment in the penthouse overloaded, and so the team would turn it off until the dish was aimed elsewhere. One day during one of these shutdowns, Ed Rawson announced, "I can hear the rep [repetition] rate!" Despite the others' skepticism, he went hunting around the penthouse. One switch box cover was 7 in. across, $\frac{1}{4}$ of the wavelength of the radar, and every time the radar pulsed, the switch cover would generate an arc. After that, the team decided not to stay in the penthouse during those tests.

Full-scale model tests have continued to be necessary during the subsequent decades of stealth research. Reducing RCS further and further requires understanding ever more subtle effects of the materials used. Scaling the structures down in size to produce a small-scale model also requires scaling up the frequencies of the radars used, so that the ratio of the wavelength to the structure size remains constant. However, although smaller structures can be built, the actual materials might respond differently at the higher frequencies. Increases in computer power have helped make simulations more accurate, but measurements of full-scale models still provide the final answer for any stealth technique.

WALLPAPER

Further development of the surface treatment was a cooperative effort between the Lockheed and Lincoln teams. The leader of the Lockheed team was Luther Duncan "L. D." MacDonald. Chemist Melvin F. George was MacDonald's main deputy. Another chemist, Perry M. Reedy, was extremely bright, and his work lived up to his middle name of Merlin. Physicist Edward Lovick had seen his boss disappear into an unknown project and learned that it was the U-2 only when he himself was brought onboard in August 1957. Other members of the team were Michael Ash and James M. Herron, who made thousands of measurements of models in the Lockheed anechoic chamber.

Lockheed personnel traveled to Massachusetts and conferred with Purcell and the others as needed. On one trip, Rodgers, Johnson, Lovick, and MacDonald met in a hotel room. To stop eavesdropping, Johnson covered the air ducts with pillows.

For security, the initial batches of the treatment were produced by the Lincoln Lab team, rather than bringing in outside contractors. Much of the testing of candidate materials happened at Edwards North Base. The heart of the test apparatus (Fig. 6) was a "magic T" waveguide, which had been invented at the MIT Radiation Lab during the war. A signal generator provided the desired frequency of electromagnetic energy. A receiver with a polar recorder measured the reflected energy and plotted it on graph paper. A horn antenna on one end of the magic T aimed the energy at the test object. Opposite the horn was a slide screw tuner, which was adjusted to give a zero indication on the receiver when no object was present. A microwave radar signal from a signal generator was fed into a waveguide and emerged from the horn antenna to illuminate the test sample, a sphere. Reflected waves returned to the horn and were measured by a receiver, which plotted the magnitude of the signal on a polar recorder. The walls on three sides were lined with radar absorbent material to prevent reflections from objects other than the test sample. So that the equipment could be left assembled in between tests, it was all mounted on what was called the "big rumbling cart."

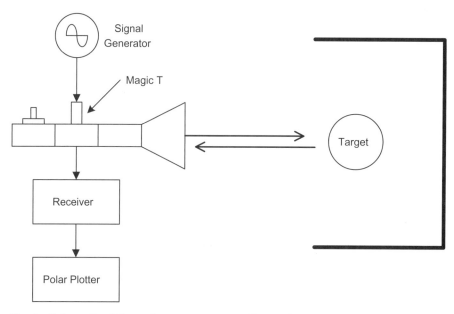

Fig. 6 Schematic of the radar test apparatus. (Drawn by the author from a sketch by Edward Lovick.)

At first the cart was aimed out the hangar door and the test specimens—metal plates treated with the material being tested—were placed in the open air in front of the radar. Later specimens were put in a drum with the slide screw tuner adjusted to eliminate the return of the empty chamber. A series of metal spheres of known diameters were measured to calibrate the radar. Johnson had an anechoic chamber built in Building 82 in Burbank, and measurements were done there, reducing the time Lockheed personnel had to spend traveling. Lockheed applied the radar-absorbent material to a cylinder they called the "barrel" for testing in the chamber. Eventually the cart was taken from Edwards to Bakersfield to measure the return from U-2s that had the treatment applied.

When the Lockheed team calculated the expected returns from their materials, they usually had to resort to slide rules. There was a Hewlett Packard desk calculator near the anechoic chamber, but they seldom were able to use it; the structural team had it tied up most of the time (Interview with Ed Lovick, Northridge, CA, 4 Feb. 2006).

Back in Massachusetts, the Lincoln Lab team continued making measurements in their penthouse. They shot the radar at models suspended in front of an open window. A number of treatments were tried and discarded. One was silver paint, which it was hoped would provide broadband protection by scattering all frequencies of radar, not just a narrow band as a Salisbury screen would. It turned out not to be effective.

The treatment they eventually chose came to be known as "Wallpaper" and also "Thermos." It was a coating for the U-2's fuselage intended to reduce reflections from S-band radars like the Fan Song. A typical S-band wavelength is 10 cm, about 4 in. A quarter-wavelength reflector as in a classical Salisbury screen would have been 1 in. thick. This would have increased drag by increasing the diameter of the fuselage by 2 in. and the surface area significantly, and it would also have added unacceptable weight and caused a loss in altitude.

To keep the weight of the coating down and to minimize its effect on the U-2's aerodynamics, Johnson insisted on a maximum thickness of $\frac{1}{4}$ in. and a maximum weight of 1 oz/ft^2. It also had to have very little aerodynamic effect (Interview with F. Robert Naka, Concord, MA, 16 Sept. 2004).

The solution was a clever implementation of a Salisbury screen. The coating consisted—from inside to out—of Fiberglass, a honeycomb spacer, a graphite-impregnated layer, a protective layer for durability, and finally a layer of paint. The graphite-impregnated layer was the key because its electronic properties caused it to behave as though it was much thicker. This meant that even though the entire coating was only $\frac{1}{16}$ wavelength in thickness, it acted as though it was $\frac{1}{4}$ of a wavelength, like a conventional Salisbury screen, while limiting the increase in drag and weight.

In its original configuration of rows and columns of squares, Wallpaper worked best at S band (2–4 GHz). Eventually it was found that by cutting a small iris in the center of each square, it would also resonate and reduce the return at X band (8–12 GHz). Although this was not a broadband solution— one that worked at all frequencies—it was still an important improvement. Figure 7 shows samples of single-frequency wallpaper.

Fig. 7 Samples of single-frequency Wallpaper (fuselage coating) recovered from the site of the crash of the U-2 prototype, Article 341: a) the coating was a total of $\frac{1}{4}$-in. thick, including the thin honeycomb; b) the outer surface was the pattern of squares. (Courtesy of Peter Merlin.)

Fig. 8 U-2 with "Wallpaper" treatment applied to its fuselage. (Courtesy of Roadrunners Internationale.)

Initially the coating was applied to the entire fuselage. However after the engineers realized that it was causing cooling problems by trapping heat, the coating was removed from the upper fuselage where the exposure to radar energy was minimal. This improved cooling but did not completely solve the problem.

When Lockheed first applied Wallpaper to a U-2, there were problems with it adhering to the skin (see Fig. 8). Back in Waltham, the Lincoln Lab team decided that the application process was wrong, but they were not exactly sure how to fix it. They had a colleague, Henry Katzenstein, briefed into the project, and he came up with a new process. At first Katzenstein's clearance did not extend to visiting the Ranch to supervise the new installation, and so the regular team members had to go.

To celebrate the initial success, Herb Miller took Rodgers, Bazemore, and Naka with their wives to the Copacabana night club in New York City. They ate so much that Pat Naka, who was pregnant at the time, gained several pounds. Her pediatrician "bawled her out," but she afterwards maintained that the party was worth it (Interview with F. Robert Naka, Concord, MA, 16 Sept. 2004).

TRAPEZE

The wings did not only produce reflections to the front and rear. When a 70-MHz radar—like the Soviet search radars of the time—illuminated the U-2 from the side, the wings produced reflections that were even stronger than the reflections from the fuselage. The solution came in stages. Probably because of the involvement of Stewart Miller, Bell Labs had been commissioned to do a number of theoretical studies, some of which involved actual measurements of the reflective patterns of metal bars. The results of their studies were given to the Lincoln Lab team, who reproduced the experiment and confirmed that the wings of the U-2 model had a similar pattern. They were able to drastically reduce the reflections from a bar by putting resistors on both ends of it. However, there was no practical way to do that with the real wing.

One day in the penthouse, Dan Schwarzkopf remarked to the antenna theorist that the reflection patterns measured by Bell Labs looked like the transmission pattern from a rhombic antenna and that he couldn't see a way to get rid of them.

> He was one of the few theoretical people I've hit who could sit at a tea table in the penthouse with a sheet of paper and tell you what to do, and in a way that I could understand. So he said, "Oh, you need a slow wave structure to slow down these currents that can't exceed the speed of light." (Interview with Daniel Schwarzkopf, Stow, Massachusetts, 30 Nov. 2003.)

The problem was that a radar wave intersecting the leading or trailing edge of the wing at an angle would induce electrical currents that would try to move faster in the metal than the speed of light. Because they couldn't, energy was reradiated in the "rhombic lobe" pattern. The slow wave structure was a resistive ladder circuit along the edges, and it would have the effect of slowing the propagation in the metal and avoiding the reradiation.

However, there would be a problem testing this. Because the 14-ft wavelength of a 70-MHz signal was too big to deal with in the small penthouse, they were using a 1/40th-scale model of a U-2. Scaling down required using a frequency 40 times higher to give a wavelength 40 times shorter. Unfortunately, normal resistors did not behave as resistors at the much higher frequency. Serendipitously, Ed Rawson showed up at that point in the conversation. He had been working with digital computer circuits and had gotten experience there with metal film resistors. At the higher frequency, these retained their resistive characteristics. So they assembled a slow wave structure on the U-2 model using metal film resistors and, "That turned out to kill the rhombics beautifully."

The slow wave structure was later adopted in a different form by Ed Lovick for the wing edges of the high-speed follow-on to the U-2. For Schwarzkopf, that moment in the penthouse where the antenna theorist came up with the solution to a difficult problem was the first major step towards a practical application of stealth research.

Kelly Johnson's initial response to the idea of stringing wires on the U-2 was to object to it. The drag of a structure in flight can be predicted if its Reynolds number is known; a low number means that the flow will be laminar (smooth), and a high number means that it will be turbulent. Turbulent flow can also disrupt the lift of a wing. Because he did not know the Reynolds number of a wire in a stream of air, he could not predict how the added wires would affect the performance and handling of the airplane (Interview with F. Robert Naka, Concord, MA, 16 Sept. 2004). However, because the anticipated benefit of a reduced RCS was so great, he provisionally accepted the change until the drag increase could be measured experimentally.

Two U-2s were dedicated to testing the RAINBOW concepts, beginning in January 1957. Article 341, the U-2 prototype, was scheduled for testing of the Wallpaper high-frequency treatment, and Article 343 was scheduled for testing of the low-frequency treatments, Trapeze and Wires [19]. Because of its appearance, the copper-plated steel wires strung along the wings and horizontal tail came to be known as "Trapeze" (see Fig. 9). Lockheed's L. D. MacDonald and Ed Lovick assembled Trapeze on the aircraft in a hanger at North Base at Edwards Air Force Base (Interview with Ed Lovick, Northridge, CA, 4 Feb. 2006). To support the wires at the ends of the wings and tail, Lockheed chemist Mel George and his group built Fiberglass bows. The intermediate supports for wires into the wind were made from sections of Fiberglass fishing poles. To adjust the impedance of the wires to form a slow wave structure, ferrite beads were strung on them at precise locations. Those beads had been calibrated in tests in the penthouse in which Bob Naka made measurements by placing them in a waveguide (Interview with F. Robert Naka, Concord, MA, 16 Sept. 2004). The values were rechecked by Lovick, who tested the completed installation in a hangar. With chordwise

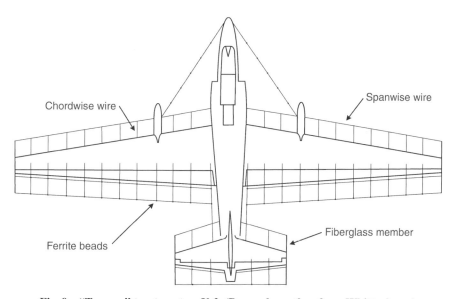

Fig. 9 "Trapeze" treatment on U-2. (Drawn by author from Whittenbury.)

wires connecting to the wing edges and ferrite beads for tuning the frequency response, Trapeze formed a slow wave structure, which induced currents that suppressed the rhombic lobes. Unfortunately, the reduction came at the cost of increased drag and reduced cruise altitude.

Because the "dirty birds," as the Trapeze-equipped U-2s would come to be known, would be seen by people not in the project, a cover story was concocted that it was an instrument for atmospheric research. Ed Lovick recalled that the wires were shiny and "... glistened in the sunlight light you wouldn't believe." As precarious as the installation looked, he only recalled one time when a wire broke in flight.

In the end, Trapeze only reduced the radar return by about 12 dB [20]. This was a magic number of sorts for RCS because it meant that the detection distance was cut in half. In other words, the return of the treated aircraft was the same as that of an untreated aircraft twice as far away. In an interview, Lovick described this as "... not sensational, but significant." (Author interview with Edward Lovick, 2 Feb. 2006.)

WIRES

At some frequencies, the fuselage generated a big reflection from the sides. The solution was to build a sort of Salisbury screen by placing dipole antennas along the fuselage and vertical tail. A dipole is a straight wire, with the length matching the frequency to be absorbed. The practical problem would be installing hundreds of dipoles, each of which would require a support at each end. That many supports would add too much weight and too much drag.

The key to solving the problem was that if enough ferrite beads were strung on the wire at one point, they would prevent current from flowing through that point. By placing groups of beads at regular distances along a long wire, the wire would behave as though it were a series of separate wires—a series of dipoles. A long wire could be strung as far along the fuselage as needed and only anchored to the fuselage where it was necessary to prevent the wire from moving in the airflow. The result was many fewer supports—and much less weight and drag—than if there had been one support per dipole.

This treatment was called, appropriately, "Wires" (see Fig. 10). In addition to the fuselage, it was also applied to the vertical stabilizer, with 12 wires on each side, running from the leading edge to the trailing edge of the rudder. Each wire had two supports, one at the leading edge and one at the back of the rudder. Another wire ran vertically just in front of the rudder (see Fig. 11). Lockheed flight-test engineer Bob Klinger recalled that the first flight testing of Wires was on a T-33 at Eglin Air Force Base, Florida, in the first week of October 1956 (Personal communication from Chris Pocock).

Fig. 10 The "Wires" treatment on the U-2 fuselage included barely-visible ferrite beads. (Courtesy of Lockheed Martin.)

Fig. 11 "Wires" installation on the vertical stabilizer and rudder. (Courtesy of Lockheed Martin.)

Some of the fuselage wires even ran above the cockpit and had to be connected after the pilot was inside and the canopy closed. Schwarzkopf recalled that "... the pilots were nervous as hell that they'd be sliced into ribbons if they ever had to bail out." (Interview with Daniel Schwarzkopf, Stow, Massachusetts, 30 Nov. 2003.)

Norm Taylor recalled a long period of trial and error:

> Each morning we drove some 60 miles into the desert, measured all day, and went back to Vegas in the evening. The desert was boring and the work tedious and frustrating. Our measurements did not seem to get better as Frank moved wires and reflectors in an attempt to reduce the reflections from the U-2 mock up. (Taylor, Norman H., unpublished memoir, 2002.)

There was also a problem because of the increased drag and weight. A treated U-2 had a maximum altitude 5000 ft less than that of an untreated U-2 and a range 20% less. None of the pilots was happy about being that much closer to Soviet aircraft. The reduced range meant that the treated aircraft could not be used on some routes.

Despite the overall performance loss, Wires actually made an improvement in some of the handling characteristics. Schwarzkopf noted that "... one of the test pilots, when we were having a drink in a Las Vegas bar, commented that we had actually improved it. Needless, Kelly Johnson wasn't going to say that." (Interview with Daniel Schwarzkopf, Stow, Massachusetts, 30 Nov. 2003.) Probably the effect was as a result of the vortices from the wires changing the airflow over the wings and control surfaces in an unanticipated way.

The final results were that 70-MHz returns were reduced 20 dB and S-band returns about 10 dB [20]. In late 1957, at a meeting in Cambridge, Rodgers presented the numbers. Schwarzkopf recalled,

> We had these polar plots of the original aircraft model, and the fixed aircraft model. And Frank very carefully said, "Okay, this is the cross section that we started with." He says, "This is what we ended up with up after our complete solution."
>
> And everyone ooh'ed and ahh'ed, and said, "Oh, boy that's marvelous." (Interview with Daniel Schwarzkopf, Stow, Massachusetts, 30 Nov. 2003.)

However, the question was still open as to whether the reduction would be enough to prevent tracking.

SECURITY

The need to maintain security often led to subterfuges by the people involved in the project. During visits to Lockheed, Johnson had some of the Lincoln Lab scientists use false names. Bazemore remembered that "... Kelly

Johnson was high on secrecy and also on drama. He came through the door and said, 'Nobody's to know who you are or why you're here. You're going to be Mr. Brown, and he points to Bob and says you're going to be Mr. Smith." Because Bob Naka was Japanese-American, it's unlikely that anyone took his cover name seriously (Interview with Thomas C. Bazemore, Santa Barbara, CA, 13 Nov. 2004).

At Lincoln Lab, there were problems with top secret documents for RAINBOW. The guards could inspect anything taken out of the building. However, they were not cleared for top secret, and they were not cleared for RAINBOW. The team members would take them to the Director's secretary, who would take them out of the building and either hand them back in the parking lot or mail them from her home. According to Dan Schwarzkopf, "She was Marshall Holloway's secretary; no one was going to ask to see her bag."

One of the test rigs used at Indian Springs Air Force Base, in Nevada, supported a U-2 model while it was illuminated by radar. The pole on which the aircraft sat was rotated by a motor scavenged from an old World War II-era radar system. The only place information on the motor could be found was in a manual in the Lincoln Lab library. However, because no one had ever bothered to declassify the radar, the manual was still rated as confidential, which was below secret. One of the scientists checked the manual out, gave it to Holloway's secretary, who gave it back to him in the parking lot, and he mailed it to Nevada.

The manual never arrived. Every year the library sent a form letter to everyone who had documents checked out. Because he couldn't risk having an investigation expose the work being done, the scientist would send back the form, saying that he still had the manual. Eventually the manual was declassified, and finally the scientist could say that he had lost it.

At Indian Springs, the pole supporting the models could be seen from the highway. Schwarzkopf remembered that "We had lunch at the little diner once and we were curious of what people thought of the pole sticking out of the ground. ... And they said, 'Oh, yeah! It comes up, rotates around, and goes down again'" (Interview with Daniel Schwarzkopf, Stow, MA, 30 Nov. 2003). The visibility of the Indian Springs site was one reason that a radar test range was eventually established at the Ranch.

He also recalled a meeting "in my hotel room with Mel George and some other people from Lockheed when like most engineers I was sketching on a piece of paper. When the discussion was over I set fire to the paper in an ash tray and then flushed the remains down the toilet. I commented that flushing the remains down the toilet was probably not necessary from a security point of view but could prevent one from getting a reputation as a firebug. Mel George laughed and said something to the effect that it was the first indication of a sense of humor he had seen" (Schwarzkopf, Daniel, e-mail to author, 16 April 2008).

FLIGHT TESTING

Living and working at the Ranch, the Lincoln Lab and Lockheed scientists and engineers came to know the test pilots well. The path and altitude of each flight had to be exact, so that the radar would measure the aircraft at just the right angle and range. This meant that the scientists and pilots had to work closely so that there were no misunderstandings about the requirements for each flight. Decades later, Frank Rodgers remembered,

> Test pilots are a breed apart. This was brought home to me rather forcefully by a couple of incidents during this period. One pilot had been on a long flight at maximum altitude in the "dirty" U-2, a name which had been given to the plane which had undergone our treatment. As he was heading home but still some 250 miles from the Ranch, he made a quick calculation which convinced him that if he cut the engine, he could glide all the rest of the way, arriving at the Ranch at just the right elevation to make a perfect dead-stick landing. So he did just that and, sure enough, his hunch proved perfect. He came in right on the money, lining up perfectly with the runway, and proceeded to land. But strangely, when his wheels should have been touching down he felt nothing. He sank a little lower before he realized he had forgotten to lower his landing gear. It was too late to do anything but pancake onto the runway which, of course, badly damaged the underside of the fuselage as well as the treatment thereon. The testing program had to be suspended while the necessary repairs were made.
>
> At one point while I was at the Ranch, a meeting was scheduled in Burbank which I was required to attend. Arrangements were made for one of the test pilots to fly me in a twin engine Cessna to Burbank, cool his heels until the meeting was over, then fly me back to the Ranch. We took off and as he put it into a steep climb he asked me whether I wanted to take the direct or scenic route. I made the wrong choice when I said, "Let's take the scenic route." No sooner were the words out of my mouth than he pushed the stick forward and dived straight toward the earth, pulling horizontal no more than ten feet off the ground. We flew at that altitude, and I kid you not, from the Ranch, over the edge of the dropoff into Death Valley, across the Valley and up the other side. We did not rise above that altitude until we approached the air traffic control zones of Greater Los Angeles. We flew for long periods along railroad tracks and when we came to telephone lines crossing the tracks, he would fly UNDER the lines. As we were climbing out of Death Valley he spotted a herd of wild donkeys which he proceeded to stampede.
>
> On the return trip, we took another route but the ground clearance was the same after we left the crowded Los Angeles area. Again we were flying along a railroad track just after sunset when he spotted a train coming toward us. He maintained his height until we were within about a football field length of the oncoming train at which time he simultaneously turned on his landing lights and pulled up just far enough to let the train pass under us.

> Later, we were flying up a dry river bed when a bridge loomed ahead of us. He asked me whether I wished to go under or over the bridge. I'll leave it to your imagination as to what my response was. (Rodgers, Franklin A., unpublished memoir, 3 Feb. 1995.)

The test pilots also were one of Tom Bazemore's biggest memories. He recalled that they made three or four times as much as the Lincoln Lab people. When they ate together, the pilots would usually order the most expensive meals; when it came time to pay, they would suggest splitting the bill equally, and so the Lincoln Lab people ended up paying more than their share. But they really did not mind. They respected the pilots for the risks they had to take because the U-2 was "terribly difficult" to fly (Interview with Thomas C. Bazemore, Santo Barbara, CA, 13 Nov. 2004). The risks were quite real. Of the 20 original U-2s built for the CIA, 12 crashed, killing nine pilots, not including the four aircraft shot down [2].

On 2 April 1957, RAINBOW suffered a fatality. Lockheed test pilot Bob Sieker was flying Article 341, the original U-2 prototype, with Wallpaper applied. The engine failed, probably because of heat buildup from the Wallpaper. As cabin pressure dropped, Sieker's pressure suit inflated, and the clasp holding his faceplate failed, allowing it to open. He quickly lost consciousness and was killed in the crash. Rodgers remembered:

> On one of the few weekends I was able to spend at home in Concord, Massachusetts, I was listening to the radio when the program was interrupted by a special bulletin. An experimental plane had crashed somewhere in the western desert and authorities were trying to find the location of the wreck. The bulletin went on to say that anyone who happened onto the site was warned not to approach the wreck even to aid the pilot if he were still alive. I knew, intuitively, that it had to be a U-2 and the shock to me was compounded by the realization that I probably knew the pilot. But I could say nothing to my wife or children and had to act as though I had not even heard the bulletin. The next day at work, I learned that the pilot was Bob Sieker, one of the test pilots assigned to fly our treated planes. (Rodgers, Franklin A., unpublished memoir, 3 Feb. 1995.)

Westinghouse Experiments

As the work expanded, Bissell brought in the Westinghouse Electric Corporation, in Baltimore, in April 1957. What was needed was a set of studies of a U-2 model. To simulate S-band radars with the 1/12th-scale model, a higher-frequency Ka-band radar was used. On 14 June 1957, he approved letter contract no. BE-2022 for "… theoretical studies, experiments and laboratory tests in the field of electromagnetic radiation, and establishment of a test range for use in connection with the studies and investigations being conducted under project RAINBOW." The initial contract was for $100,000 in FY 1957 AQUATONE funds [21] and was quickly increased to $150,000 [22]. The contract was for work that had actually begun on 1 March 1957 and was planned to continue until 30 June 1958.

In October, the Agency authorized Westinghouse to build a 2400-ft antenna range on their property; some of the work was done by the Baltimore firm Kirby and McGuire [23]. By December they were ready to emplace the antenna, and the Big Boy Rigging Company provided a crane for the sum of $245.68 [24]. Three weeks later, the letter contracts were replaced with a definitive contract to cover the $150,000 [25]. At first use of the range was provided to the Agency under a lease arrangement, but that was cancelled on 10 February, and the Agency transferred title to the poles on the range to Westinghouse in exchange for being released from a requirement to restore the site [26]. Measurement work continued into early 1959, with the range being used at least part time by Edgerton, Germeshausen, and Grier (EG&G) [27].

The range had originally been intended to be large enough to place the radar 4500 ft from the target, but there was too much interference from nearby objects—clutter. Eventually, the distance was cut down to 600 ft, which was almost so close that the curvature of the spherical front of the radar wave was so tight that it would strike different parts of the model at different times and induce unrealistic reflections. At this range, though, the magnitude of the clutter reflections was reduced to 15 dB below the model reflections, which was enough for valid measurements.

The study identified six major reflections that were fairly constant over a range of elevation angles from 10 to 45 deg. One was a narrow reflection from straight ahead, one perpendicular to each wing, a wide reflection from each side, and a narrow reflection from the rear. To determine whether a particular structure of the U-2 was the source of an echo, it would be covered with 2-in.-thick horsehair batts and the return measured for comparison with measurements made without the batts. The batts would reduce the echoes down to the level of the background clutter. In this way it was determined, for example, that the flat wing-tip skids were a major contributor to the side reflections.

Figure 12 shows strip chart recordings of the strength of radar reflections from a U-2 model. Polar (circular) charts of reflections proved to be much easier to interpret.

The next step was to try to reduce the returns by applying sheets of scaled-down absorber material to the areas causing the largest returns. This turned out not to give consistent results, and Westinghouse concluded that either the material was not working as expected or the instructions they had been given on using it were wrong.

To understand what was wrong with the use of the absorbers, they set up a small 10-ft range and began a series of simplified experiments on a variety of targets, 6-in.-long aluminum cylinders ranging in diameter from 1.3 to 6.5 in., as well as 6-in.-square aluminum sheets. Once the basic measurements

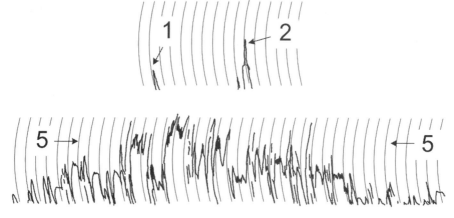

Fig. 12 Westinghouse made strip chart recordings of the strength of radar reflections from a U-2 model. As the model was rotated about its vertical axis, the recorder marked the paper strip with the magnitude of the reflection: (1) was the reflection from directly off the nose and (2) was when the beam struck perpendicular to the wing leading edge; the area (5) was the large reflection from the fuselage and possibly included the rhombic lobes from the wings.

were done, they then applied the absorber material to the cylinders and sheets.

One result was that the absorber worked best for the flat sheet, less well for the largest cylinder, and worst for the smallest cylinder. The effectiveness varied greatly between sheets, indicating quality control problems in its production. They also found that if the absorber had a sharp edge parallel to the axis of the cylinder, the reflection was even worse than for a completely bare cylinder. The transition would need a gradual reduction in absorber thickness over an arc of about 40 deg. By cutting the absorber edge in a serrated pattern, the echo fluctuations along the transition were reduced to almost nothing. By adjusting the valley of the serrations to be 30 deg below the horizontal line of the fuselage and the peaks 10 deg above—which assumed that the U-2 would first be detected at an altitude of 70,000 ft and a range of 60 n miles from the Soviet radar—then the reflections would decrease as it got closer and the radar looked up at a larger angle.

They also learned that two sheets of absorber should not be butted up against each other, even with a very small gap. Instead, each should be cut in matching serrated patterns and the peaks of one side's serrations placed in the valleys of the other side's scallops [28].

RAINBOW MISSIONS

Not all of the necessary data could be obtained by measuring the U-2. To gather information about the locations of Soviet radars, the Agency commissioned the Ramo-Wooldridge corporation to develop System 5, a multiband receiver that fit in a conventional U-2's Q bay and connected to an antenna mounted on the bottom of the fuselage. Not only would the data help mission planners to avoid radars, but they would also give clues to unexpected radar frequencies. On 18 March 1957, Jim Cherbonneaux flew the first mission with a System 5 along the Soviet border with Afghanistan [2].

On 6 May, Bissell and Dulles briefed the President on AQUATONE, including the expected results of the RAINBOW work. In his prebriefing notes, Bissell wrote,

> It is believed that the radar reflectivity of the U-2 aircraft can be so reduced as to create a good chance that a majority of overflight missions will avoid detection entirely. Nevertheless, it must be anticipated that at least a certain proportion of them will be detected, although their continuous tracking should be extremely difficult [29].

He added that if the radar camouflage were effective, then not only would detection and tracking be difficult, but also interception, even after the Soviets had developed aircraft or missiles capable of reaching the U-2's altitude [29].

In the U-2 operational detachments, the dirty birds were known as "Covered Wagons." One had been sent to each of the detachments. The first two flights were made by Jim Cherbonneaux on 21 and 30 July 1957, out of Turkey. The System 5 receiver in the aircraft showed that the Soviet radars were alerted when the U-2 was flying directly toward or directly away from the radars. The conclusion was that the source of the returns were the exhaust, the inlets, and the cockpit, all areas that could not be treated.

James Reber, Bissell's intelligence requirements chief, realized that signals intelligence reports that were distributed outside the CIA might reveal the existence of RAINBOW to readers of the reports. In Germany, the 6901st Special Communications Group monitored Soviet bloc reactions to western reconnaissance flights; the 6902nd, based in Japan, served the same purpose for the eastern USSR and China. The units would issue a daily mission report (DMR) with their findings to both the CIA, the National Security Agency (NSA), and other customers. The DMRs were then used by the NSA in the preparation of a more in-depth report known as a CMR (probably an abbreviation of consolidated mission report) (Donoghue, Joseph, e-mail to author, 9 Oct. 2007). The Air Force officers who prepared the CMRs were not cleared for RAINBOW, but they might be able to infer the use of some techniques to defeat radar tracking.

The solution was to clear the officers who prepared the CMRs and to restrict the distribution of DMRs and CMRs that reported on RAINBOW flights. A limited number of recipients of the reports were also cleared, including people in Naval intelligence and in Army intelligence. In a bit of disinformation, Reber would explain the new restrictions by blaming the NSA [30]. A month later, Jack Gibbs approved distributing the CMR to one more supervisor at

NSA, saying that "I doubt if anyone not alerted to Rainbow would even notice any unusual information being transmitted or withheld" [31].

Operations continued with both untreated and treated U-2s. On 5 August, Eugene "Buster" Edens flew a covered wagon to the Aral Sea and back without any apparent detection; the lack of tracking was ambiguous because radar coverage was known to be weak in the area. On 12 August Bill McMurray flew 1400 miles along the Sino–Soviet border with no sign of tracking. However, on 28 August, L. K. Jones had to turn back on a flight toward Stalingrad. Because of the drag of the treatment, he could not get above 66,000 ft, and he estimated that the Soviet fighters following him were only 1000 ft below [2].

On 23 August, Bissell, Cabell, and Twining met with the President to report the results of the flights to date [32].

THE SHIP IS STILL SINKING

Although the Covered Wagon flights would continue into the following spring, by August 1957 it was becoming obvious to Frank Rodgers and the others that they had done the best that they could with the U-2:

> After almost a year of work on the U-2 we had accomplished about as much as I felt possible in reducing the cross section of the aircraft and I so informed Bissell. It was not enough to put Eisenhower at ease, particularly when considered in conjunction with the price paid in aerodynamic performance. Bissell's response: "You're telling me the ship is still sinking but maybe more slowly?" I laughed. He then asked me, "If you could influence the design of an aircraft from its inception rather than working a quick-fix on an existing plane, do you think you would be any more successful?" I replied that it was possible but again I could guarantee nothing. He asked me to turn my efforts in that direction. (Rodgers, Franklin A., unpublished memoir, 3 Feb. 1995.)

PASSIVE ECM COMMITTEE

On 6 September 1957, the Lincoln Lab team had a chance to compare their work on electronic countermeasures (ECM) with that of the Defense Department. The meeting is interesting both for what was said and what was not said. George Valley, associate director of Lincoln headed the committee, which was attended by Air Force Colonels Appold, Nunziato, and Lewis, a Commander Peterson and a Doctor Wright of the Navy, and by Rodgers, Bazemore, and Butman, from Lincoln.

The Air Force and, to a lesser extent, the Navy were funding a variety of radar camouflage development efforts by both industry and academia, in both the U.S. and Europe. The Air Force presented work at seven different institutions. Gaetano Latmiral, at the University of Naples, was trying to derive the properties of a material that would give a 20 dB reduction at X band. The University of

Goettingen was studying resonant and magnetic dipole absorbers. Brooklyn Polytechnic was studying lossless dielectrics and resistive sheets for layered absorbers. Bjorksten Research Laboratories, in Madison, Wisconsin, were looking at magnetic materials in a dielectric matrix. Deutsche Magnesit, in Munich, had developed a multilayer painted-on absorber; with a thickness of only 2 mm, tests showed 95% absorption at 9.3 GHz (X band). Battelle Memorial Institute was also trying to find ways to reduce the RCS of aircraft at lower frequencies.

Emerson and Cuming seem to have made the most progress, having developed an absorbent material embedded in a plastic honeycomb. A 1-in.-thick layer of this material was showing 95% absorption at frequencies from 2.5 to 13 GHz (S and X bands). This was so promising that it was planned for application to a T-33 for flight tests, in a program that became known a PASSPORT VISA. (The PASSPORT VISA flight tests eventually began in late 1958. One of the test pilots was Virgil I. "Gus" Grissom, later a Mercury, Gemini, and Apollo astronaut and the second American in space [33].)

Rufus Wright, of the Naval Research Laboratory (NRL), described efforts to reduce the RCS of naval vessels. Perhaps because he did not describe any aircraft applications and the note taker at the meeting was an Air Force major, Wright's presentation only rated two sentences in the meeting minutes. In fact, Wright and his NRL team, working with Emerson and Cuming, had actually done pioneering work in graded dielectric materials. Eventually the Navy had ended funding, and Emerson and Cuming obtained Air Force funding, developing the material reported at the meeting [33].

Rodgers then presented the work being done at Lincoln Lab. Although he used the project name of RAINBOW (and warned the participants that the name was confidential), he did not mention the CIA. He described it as an effort between Lincoln and Lockheed, with Polaroid developing the coating and Westinghouse performing model measurements. He described how they were trying to replace the original concept of a sheet of conducting squares with a homogeneous coating.

He then described some of the practical problems Some were with making measurements, such as Westinghouse's having only a 15-dB range in their equipment; if the reduction were greater than that, then the equipment would only see noise, and no useful number would be obtained. Lockheed had apparently been using an application technique that was interfering with the electrical properties of the coating; the lesson was that the material could not be developed independently of the application process.

He apparently did not mention any plans for a new aircraft [34].

Flying Saucer

Frank Rodgers would disappear for days at a time while he thought through problems. Once, after a four-day absence, he came in with an idea for an experiment. He gave a coffee saucer to one of the engineers and told him to cover it with metal foil and then measure the radar return. The reflection was significantly below that of a similarly sized sphere and a great improvement on the reductions attained on the U-2 using Wires and Trapeze. Norm Taylor asked him where his idea came from. He replied, "Viewed from below, all reflections go away from the source." (Taylor, Norman H., unpublished memoir, 2002.)

The insight was the culmination of a long effort that had begun with Bissell's suggestion to consider starting from scratch on a new airplane. Rodgers later described it:

> For the next month I spent most of my time alone in the shack on the roof of the Lincoln Laboratories building where our experimental program had begun. There, I returned to basic research, trying to understand the nature of the radar return from simple geometrical shapes without regard to the aerodynamic practicality of such shapes. My favorite shape was a simple circular aluminum disk. It had the advantage that the pattern of radar return from it was simpler than any shape other than a sphere. I proceeded to compare the measured return from the bare metal disk to that measured when the disk was treated in various ways. After almost a month of frustrating failure I adapted to my metal disk a technique which had been developed during World War Two for an entirely different configuration. To my complete surprise, the radar return did not just decrease, it disappeared!
>
> After exhaustive checks to assure myself that my equipment had not failed, I sat back to contemplate the significance of my discovery. I realized that this treatment was what radar engineers call a "broad-band" treatment—that is, it was applicable to any radar operating at any frequency within a very broad range of frequencies. In contrast, the techniques we had applied to the U-2 were "narrow-band." If a radar were encountered, operating at a frequency different from that for which the treatment was designed, the effectiveness of the treatment could be significantly reduced.
>
> Unfortunately, a flat disk would be about as aerodynamically stable as a circular Persian rug. Its stability could be increased somewhat by distorting it into the shape of a frisbee or by fattening it in the middle so that it resembled a flying saucer. In fact, I half-seriously considered the possibility that flying saucers of the day were so hard to detect on radar precisely because extraterrestrials had already applied my discovery to their spacecraft. (Rodgers, Franklin A., unpublished memoir, 3 Feb. 1995.)

The World War II-era technique that Rodgers had adapted was layering circular sheets of Teledeltos paper (a paper with a constant resistivity) on top of the metal disc. The first sheet was the largest diameter, and each successive sheet was a smaller diameter. By the sixth sheet, the resistivity was down to 300 Ω per square, the same resistivity as the free space through which the radar wave was moving, which would prevent reflections. If they could get rid of wings and anything else that would produce rhombic lobes, then they could get the cross section way down (Interview with Daniel Schwarzkopf, Stow, MA, 30 Nov. 2003). Eventually Rodgers realized that a disk that was thick in the center and thin at the edges—a saucer—would have a similar effect, and the coffee saucer experiment resulted.

Bissell was extremely impressed with this discovery and sent Rodgers to tell Kelly Johnson. Norm Taylor went along; it was his first visit to the Skunk Works. He saw that

> Johnson was an unusual VP. Instead of the usual deluxe office he had a full scale drafting board in the center of his office, and the engineers came in ad lib to pick up the latest detail as it would come off the board. His method was awarded recognition for reducing the time for a first model by many months—fifteen to twenty, compared to three to four years. (Taylor, Norman H., unpublished memoir, 2002.)

Frank started to explain the principle to Kelly. He said that the ideal shape for a stealthy aircraft was a flying saucer. Kelly replied with "For Christ's sake!" and said that he couldn't make a real aircraft with that shape, adding that Rodgers obviously knew nothing about aerodynamics. Rodgers, who could be just as headstrong as Johnson, was unintimidated and told Johnson that he knew nothing about radar. That was the end of the meeting.

Bissell told them not to give up, but to educate Johnson. Rodgers and Taylor decided that "… it wasn't our idea that was wrong, but our presentation." They decided that they couldn't tell Johnson what the final airplane should look like, but they could give him guidelines about how to reduce the RCS and let him design using those rules. "We finally reduced the specifications to a simpler direction: to build an airplane with no flat surfaces, no straight-lines, and no concave surfaces" (Taylor, Norman H., unpublished memoir, 2002).

Johnson had the Lockheed radar team test a variety of lenticular (lens-like) shapes and confirmed that the RCS properties were good. Aerodynamically, however, they proved to be too unstable, and Johnson said that he wished that he knew how to control them. Had he been able to devise a control mechanism, he would have pursued the saucer as the basis of a serious design (Interview with Ed Lovick, Northridge, CA, 4 Feb. 2006). Nevertheless, some aspects of the saucer could still be incorporated in more conventional designs.

Scientific Engineering Institute

In October 1957, the RAINBOW work was moved out of Lincoln Laboratory. For some time there had been objections to its presence because the members of the steering committee were not cleared for it. In particular,

Jerome Friedman, head of the Radar Division and Frank Rodgers's boss, felt that work for the CIA was inappropriate. In the same week that Sputnik was launched, the Scientific Engineering Institute (SEI) was incorporated, with offices on the second floor of a bank building on Main Street in Cambridge, Massachusetts. The company's labs were several blocks away, on Binney Street. Richard Leghorn served as its first president.

Ostensibly, SEI's mission was to serve as a liaison between the Agency and the numerous industrial and academic researchers in the greater Boston area. In fact, its first activity was to serve as a conduit for funds from the Agency to Leghorn's new company, Itek, which was developing the cameras for CORONA, the first spy satellites. Although Leghorn participated in some discussions of RCS reduction, he was too busy with Itek to become deeply involved in RAINBOW (Telephone interview with Richard Leghorn, 2 July 2004).

Over the next several months, SEI became the home for most of the RAINBOW scientists, including Frank Rodgers, Ed Rawson, Norm Taylor, Jay Lawson, Dan Schwarzkopf, and others, as well as new hires such as electrical engineer Brint Ferguson and machinist Fred Bevins. Chuck Corderman and Joe Klein, who had never worked on the project at Lincoln, were hired by SEI. Some, on the other hand, elected to remain with Lincoln Lab. Tom Bazemore had been skeptical from the beginning about the practicality of trying to make an existing aircraft stealthy. When an opportunity arose to join a disarmament conference, he left the project. Similarly, Bob Butman went on to other work at the Lab. Bob Naka never joined SEI and had no further contact until March 1961, when he began work on the follow-on aircraft. And the antenna theorist who had made such a large contribution stayed at Lincoln Lab.

By April 1958, Leghorn had moved on from SEI to devote his full time to Itek. (Because Leghorn had won a patent suit with Eastman Kodak, the joke in the office was that Itek stood for "I Took Eastman Kodak.") Norm Taylor had become president of SEI, while Frank Rodgers remained the technical director. Charles F. "Chick" Brown, the associate general counsel for the CIA, became the vice president and treasurer.

Some of the team members would often discuss work in low tones while they ate lunch or dinner in a restaurant downstairs from the SEI offices. As new ideas would come up, they would sketch designs and write formulas on paper napkins, and then gather all of the napkins and carry them away at the end of the meal. One day they had paid and left the building when their waitress ran up to them. She was bringing a marked-up napkin that they had forgotten. Other times, before the team grew too large, they would all climb into a large Mercedes sedan belonging to Bevins and drive to the Durgin Park restaurant in the Boston market district.

The Binney Street lab was a rather dark and dingy one-story warehouse. On his first visit to the lab, Ed Lovick was escorted by Herb Miller. Lovick's

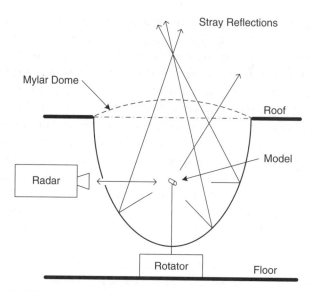

Fig. 13 Diagram of the "bathtub" chamber in SEI's Binney Street lab.

impression was of a movie set for a gangster hangout where bad things happened.

At Binney Street it was impossible to aim radars out the window, as they had done in the shack on the roof of B Building—there were too many other buildings around to reflect the beams back and create interference. In principle, an anechoic chamber could have been used to absorb the stray emissions. However, high-quality radar absorbent material to line the chamber was not easily available at a reasonable price in those days. It would be simpler if they could cut a hole in the roof, cover it with a material that was transparent to radar frequencies, and send the stray emissions through the hole.

Frank Rodgers solved the problem of directing the emissions upwards by devising a device called the "bathtub" (see Fig. 13). The bathtub was the lower half of an ellipsoid that was about 15 ft across at its widest diameter. It used the principle that any line drawn through one focus of an ellipsoid when reflected off the inside surface of the ellipsoid will pass through the other focus. The bathtub was lined with reflective material, and the model under study was placed at the lower focus, supported by a vertical pole that was rotated by a motor. The radar unit transmitted its beam in through a small hole in the side of the bathtub. Any stray reflections that missed the model or that came off the model but didn't go directly back out the hole would be reflected off the inside of the bathtub and up through the mylar window. Part of the back of the bathtub would open to allow changing the model on the rotator (Schwarzkopf, Daniel, e-mail to author, 16 April 2008).

The problem with the setup was that only very small models—about 12 to 16 in. across—could be used, limiting the detail which could be studied.

Moreover, the equipment had a fairly limited frequency range (Interview with Ed Lovick, Northridge, CA, 4 Feb. 2006). Nevertheless, it allowed the team to make a number of advances. The dual-band (S and X) version of Wallpaper was developed at Binney Street. Other treatments that were considered for the follow-on aircraft received extensive testing there. Much of the experimental work using the bathtub was done by Brint Ferguson and Joe Klein.

With guidance from Ed Rawson, Ferguson developed an important piece of equipment. It was a logarithmic amplifier, which was need to measure signals over a wide range of received signal strengths. It was important for several different applications, such as comparing the reflection from an untreated surface with that from a treatment which gave a large reduction.

Besides the bathtub, the back area of the building had enough room for everyone to park their cars. A heated garage was a considerable perk during the Boston winters.

Chapter 3

Rainbow Phase II

The results of the first Dirty Bird flights gave Bissell's team the opportunity to evaluate the success of the first year of Project RAINBOW and to plan the next steps. At the same time, the Agency (CIA) began to share what they had learned with the defense community and to look outside for additional ideas. They had also been conducting their own studies of the tracking and interception problem.

Interception Study

By late September 1957, the Agency had compiled seven months of new data on Soviet interception capabilities and issued a new study that had updated a study from the previous February. Signed by Herbert Scoville, the assistant director for scientific intelligence, it looked at radar warning and tracking, ground-controlled interception, aircraft capabilities, and missile capabilities. For most of the Soviet Union, including any areas where interesting reconnaissance targets were expected, the probability of detection ranged from high to certain. The only areas with a low probability were on the southern border east of Tashkent, east of the Ural Sea to Semipalatinsk and south of the Trans-Siberian Railroad, and east of the Ural Mountains from the Trans-Siberian Railroad north to the Arctic coast.

At the beginning of the year, the Soviets had been estimated to be fully capable of running ground-controlled-intercept (GCI) operations, and the evidence since then had strengthened the assessment. ROCK CAKE height-finding radars had been added "fairly widely" in covered areas, so that the Soviets would know the altitude of the U-2 flights. Also, radars in interceptors were believed to be effective.

One of the possibilities considered and rejected in February was that the Soviets would either develop a rocket-boosted aircraft or adapt a high-altitude research aircraft for interception. These were still deemed to be unlikely. However, four aircraft were considered candidates for being modified for interceptions at 70,000 ft: the MiG-17, MiG-19, Su-17, and MiG Ye-2A. All were thought to have service ceilings between 59,000 and 61,000 ft. The study assumed that if a crash program had been started at the time of the first

U-2 overflights, then by July 1957 the MiG-19 could have been modified to fly at 68,000 to 70,000 ft. The main threat throughout 1958 was felt to be such a modified aircraft using cannons or air-to-air missiles.

Of lesser danger were using existing aircraft to launch air-to-air missiles in a climb, using supersonic aircraft in a zoom climb, and using nuclear-armed air-to-air missiles. The GCI and missile guidance problems were felt to be too complex to allow more than a minimal chance of success. While time was running out, the chances of interception was expected to be low through the end of 1958.

Surface-to-air missiles (SAM) were not expected to be much of a danger before the middle or end of 1958. However, an advanced SAM was thought to be under development at Kapustin Yar, and that in 1959 it would be deployed around Moscow and Leningrad.

To counter these expected improvements in defense, Scoville recommended operational changes in U-2 missions, such as foregoing some targets, making frequent course changes, and careful selection of routes to and from targets [35].

The first discussions of a new aircraft were held in Cambridge on 3 October 1957. The design and designer is a mystery because the only mention is of a model that was shown to the meeting participants. Westinghouse was tasked to develop a similar model and perform a series of measurements on it.

On 14 November, Bissell and Gibbs visited Air Research and Development Command (ARDC) Headquarters, in Baltimore, to brief the senior staff about AQUATONE and plans for future developments, especially in the area of RCS reduction. They spoke with the ARDC commander, Lt. Gen. Samuel Anderson, his deputy, and two one-star generals, Marvin Demler and Don Flickinger. Bissell had previously exchanged information with ARDC by way of Col. Ralph Nunziato. Now he was trying for a tighter relationship to exchange information on the latest developments. The result was that ARDC's Maj. Gerald White, who had previously been cleared for AQUATONE and RAINBOW, was assigned to be a liaison officer to the Agency [36].

NEW RECONNAISSANCE SYSTEM

On 19 November, Bissell wrote a memo to Secretary of the Air Force Donald Quarles laying out his overall plan for the next step in the project. A recent semi-annual meeting of the President's Board of Consultants on Foreign Intelligence had recommended "... that an early review be made of new developments in advanced reconnaissance systems" [37]. They were referring to both DoD work on reconnaissance satellites and to the Agency's work on a follow-on to the U-2. The memo was input for a joint response by the DCI and the Secretary of Defense to any questions that might be asked. If the response was not handled correctly, then too much of the project could be exposed, and its hoped-for advantage over the Soviet defense system could be lost.

Bissell explained that the original intent was that Project RAINBOW would permit some reconnaissance missions to go undetected by the Soviets and to reduce the accuracy and extent of tracking of those that were detected. However, by midsummer 1957 they had realized that applying passive camouflage to a conventional aircraft—such as the U-2—would have very limited success. The main problems were that all known devices were either too heavy or bulky to apply to an aircraft without significant impact on performance, or they were inherently narrowbanded. At the same time, the number of frequencies used by the Soviets was expanding, and so any protection against two or three frequency bands would not be enough.

To provide protection in existing and future frequency bands would require a "... radical approach [which] would involve the use of unconventional materials, unconventional structures, or unconventional configurations of aircraft ..." and that would mean "... an entirely new aircraft optimized with respect to radar reflectivity."

Bissell then went on to summarize the results of the past three months of studies. At this point the project was trying to invent new RCS reduction techniques, as well as measurement techniques to evaluate their effectiveness. They had already recognized that there would be a penalty in aircraft performance, and so "... recent advances in the state of the art in aerodynamics must be reviewed in an effort to offset as far as possible the inevitable penalties. ..." In the next phase of the work, he expected to be able to quantify the RCS-vs-performance tradeoff for each technique and then focus on the one or two or three most promising ones.

He explained that the core of the technical staff was the recruits from Lincoln Lab, and that they would be supplemented with two or three more electronics specialists and assisted by consultants from other organizations. Tests by other firms would continue, with indoor low-frequency tests against small models and outdoor high-frequency tests against actual aircraft—on the ground and in flight—would use facilities already established for RAINBOW.

Bissell then described how the Air Force and Navy had been made aware of the early work on RAINBOW, and when the emphasis shifted to a new aircraft, a few senior people were informed. He said that the bulk of the government's competence in radar research was in those two services, and that the CIA did not have a parallel capability and should not try to develop one. The result was that "the most intimate cooperation" between the Agency and the Air Force and Navy would be needed.

Bissell stated that if a stealthy vehicle proved to be feasible, then it would have an advantage over a nonstealthy one that was faster or higher flying. He knew that the successor would not be operational before the spring of 1959, whereas the U-2 would be obsolete and the need for reconnaissance even greater. To that end, the management and organization of the project would require the utmost speed. He admitted that there were not a few crucial secrets, and that the Soviets knew all of the basic principles and were as competent as the Americans in figuring out the design approaches. Thus, the "... only way to achieve a decisive lead over their radar defense is to develop a system and have it operational before they have discovered that an intensive effort is being made in this area."

Finally, he laid out the plan of action. The current research would continue with the objective of selecting the design approach within three months. The Agency would contact manufacturers concerning unconventional materials and structures and would tighten their liaison with the Air Force and Navy. At the same time, they would try to limit discussions with manufacturers and to avoid issuing formal requirements that would stimulate unusual interest in the concept of a stealthy airplane. Once an acceptable design was found, they would proceed with a crash program and produce eight to twelve aircraft.

Bissell recommended that the response to the Foreign Intelligence Board totally avoid any description of the concept, and simply say that a system was being studied with great urgency, that funding and management were under control, and that if the system proved feasible then there would be a recommendation for action [37].

Basic Stealth Techniques

By the end of November 1957, Bissell had learned enough that he could lay out promising techniques for reducing RCS as well as a series of experiments to narrow down the choices.

There were four basic techniques. First, to cope with high frequencies, the exterior of the vehicle could be made completely reflecting and the radiation reflected at "innocent angles," that is, away from radar receivers. Second, radiation could be absorbed using either layers of materials or by external dipoles like the Trapeze and Wires installations on the U-2 dirty birds. A third technique applicable to low frequencies was to absorb the energy by "softening" edges by using materials having a graded conductivity. Finally, there was the use of radar-transparent structures that would reflect very little. In other words, the choices for dealing with radar energy would be to reflect it away, absorb it, or pass it through.

Specular reflection—a bright reflection back at the radar transmitter—could be calculated accurately and totally avoided if the aircraft had no concave corners. The unknowns about a reflective aircraft were then nonspecular reflections (such as from sharp convex corners) and what compromises would be needed to produce an aerodynamically acceptable design. Bissell hoped that measurements on models could answer these questions.

To reduce detection by X-, S-, and L-band radars, it seemed likely that a combination of reflection and absorption would be needed. Shapes with sharp angles would be inefficient aerodynamically, and if the angles were smoothed out, then absorptive materials would be needed. The existence of lightweight broadband reflective materials was uncertain. If the materials turned out to be bulky, they might be acceptable; if they were heavy, they would not. One possibility would be to use nonconductive materials like Fiberglass as the skin of the aircraft, with the absorptive materials inside to hide reflective structures like the engine. An alternate structure would use a very thin nonconducting skin—like Fiberglass or Mylar—but without absorptive materials directly under it; Ed Purcell had nicknamed this a "squid" structure.

Bissell recognized that the softening technique to achieve broadband absorption needed a lot of research. It was being most seriously considered for wings, where the softening could be achieved by applying materials that extended inward from the leading and trailing edges. This was less practical for the fuselage because material as much as 3 ft thick might be needed to absorb low-frequency radiation. Spiral antennas were seen as one possible solution for wing edges.

The fuselage was seen as the most intractable problem. Work was proceeding on improving the Wires technique of the U-2, dipole antennas parallel to the fuselage. Although a 12-dB reduction had been achieved, results from operational flights with the dirty birds indicated that this was not enough.

Bissell then laid out experiments that were needed and their status:

1) First would be S-band measurements of 1/10th-scale models of highly reflective designs: Westinghouse was making measurements of a rough scale model based on a design shown at the 3 October meeting, with a report due on 4 December. Bissell suggested also measuring a model of a high-aspect-ratio flying wing with a plastic empennage.

2) Second would be X-, S-, and L-band measurements of absorptive materials up to 1 ft thick: He was concerned that he had not seen much progress on lightweight absorbents and thought that people were assuming that they would be easy and thus had not actually tried to obtain any. He was not convinced that they were available.

3) Third would be calculation and measurement of plastic structures of various thicknesses at different frequencies: He had been told that this was calculable and no testing would be needed. The numbers given were that a plastic aircraft would have a return 20dB less that a metal one of the same size, and he was skeptical. He was concerned that they would need to combine reflective with transparent materials.

4) Fourth would be measurement and high and low frequencies of a scale model of a simulated wing with central metal panels and softened edges: He felt that although the work in Cambridge was very interesting, the staff was

still spending too much time trying to perfect the technique theoretically and needed to make some actual measurements on the designs that they had.

5) Last would be model tests of parallel antennas at low frequencies: He felt that this could be tested in a few days at Lincoln, and that other techniques for low-frequency absorption needed to be investigated; fortunately, one of the staff was beginning to work on this [38].

On 2 December, Jack Gibbs visited ARDC for discussions on subjects relevant to both the U-2 and the follow-on. General Demler had been briefed on plastics by his most knowledgeable staff member and told Gibbs that the state-of-the-art work was being done by Narmco, whom he described as "... probably one of the most flexible, aggressive and alert to new R&D techniques. ..." Although Zenith had more production experience in the past, they had fallen behind in basic research in fabricating techniques and forming and were showing very bad quality control and performance on a project with Fairchild. Demler suggested that if 3M, which had recently purchased Zenith, was aware of a definite application for plastics, then it might shake up Zenith, "... fire a few people, and get the show on the road." Gibbs noted in his trip report that when he and Bissell visited the West Coast, they should visit both Narmco and Zenith.

Demler and Gibbs also discussed keeping the Ranch facility open, at least in a stand-by mode. Gibbs proposed to put the requirement to use the facility in writing to Col. Geary to be passed to Air Force Headquarters.

Toward the end of the visit, there was a minor security breach. Colonel C. H. Lewis, chief of the Aero-Electronics Branch at ARDC, was present in Demler's office; Demler assured Gibbs that Lewis was fully cleared on the Agency's work. Gibbs then proceeded to describe the Agency's activities and plans for a follow-on to the U-2. Afterwards, Maj. White took Gibbs aside and said that Demler apparently was not aware that knowledge of the plans for a follow-on was more restricted and that Lewis was not in fact cleared. Back at the office, Gibbs noted that the Security Officer should contact all people cleared for the follow-on and remind them that it was a separate project and that not all AQUATONE- or even RAINBOW-cleared people could be told about it [39]. This reduction in the number of "witting" people would happen at each major step in the project, a classical use of the principle of "need to know."

SUMMING UP

On 4 December, Bissell held a meeting at which the technical staff presented their recent findings. The four basic techniques had been somewhat refined. Reflection could be done by either the external surface or by a partially buried shield; an unconventional configuration of the aircraft was unavoidable. Absorption of higher frequencies could be done by a 1- to 3-in. layer of lightweight material. Low frequencies could be absorbed by wing edges treated to soften them electrically. Transparent materials could work at all frequencies, and nonconductors were now known to be transparent at low frequencies and could be used for structural members. To protect them from high frequencies, they could be buried in foam of graduated density, which would avoid an electrical discontinuity that would create a reflection. Absorption of low frequencies by dipoles as on the U-2 was now known to be inadequate.

The flying saucer appeared to be the most significant new discovery. Because it provided broadband protection, Bissell saw it as "enormously promising." The shape of the vehicle did not have to be a saucer; a saucer-shaped shield could be a thin layer inside a nonconducting structure or painted on a frame under the aircraft's outer surface. The team was busily investigating related shapes, such as ellipsoids and multiple discs.

The team had examined the use of plastics and knew that their usefulness was limited because they were reflective, although not as much as metals. At S band, the return from the surface of a plastic aircraft would probably be not better than 15 dB less that that of a metal airplane. And that ignored the problem of preventing reflections of metal structures inside the plastic shape. At X band, there was less difference between the reflections from plastic and metal shapes.

They had worked out a number of design principles:

1) The only way to protect engines and other metal structures would be to use discs or other reflective shields.

2) Because the entire aircraft could not be a disc, there would be structural members exposed to low-frequency radiation. The solution would have to be making them transparent by using plastics and eliminating metal components inside them.

3) If a disc- or saucer-shaped shield were buried inside a structure, S- and especially X-band radiation would be reflected from the outer structure. That meant that the outer structure, even if made of plastic, would have to be shaped for reflection.

4) Because the wings and empennage could not be shielded, they would also have to be shaped for reflection at innocent angles.

5) Even though broadband "innocent reflection" seemed to be the primary technique, softening of edges would still be needed.

Finally, the team decided absorption of high frequencies would not be a primary technique. The materials that could do the job either would cause reflections at low frequencies or would be unnecessary because of the use of reflection [40].

The meeting resulted in a list of additional experiments to perform (see notes in Fig. 14). One was to measure the effects of removing wires parallel to the front and leading edges of the wings. A second was to modify an S-band

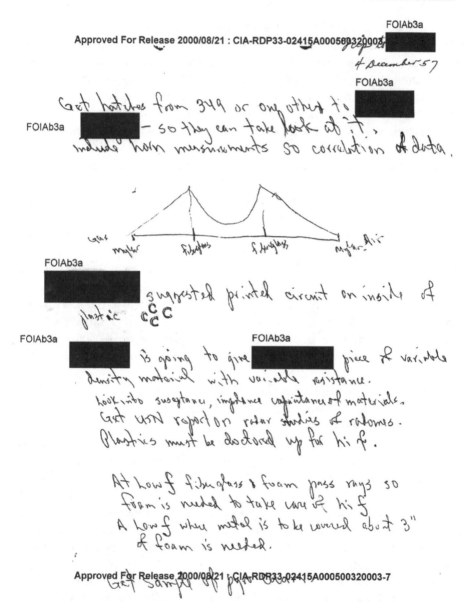

Fig. 14 Notes by Col. Jack Gibbs from the 4 December 1957 meeting. (Courtesy of the Central Intelligence Agency via the National Archives and Records Administration.)

technique for a broader frequency band by using current materials with a thicker base and to extend it to X-band protection by using a double layer of the material. Third was to move ahead on measurements of edge-softening techniques. The staff in Cambridge would attack the first three items. Finally, Project Headquarters would contact NACA to learn the feasibility of a structural foam that could be stressed both in tension and compression [41].

On 10 December, Bissell visited yet another plastics manufacturer to learn their capabilities for research, development, and fabrication of Fiberglass parts for the follow-on aircraft. The company (which was apparently neither Emerson and Cumming nor Owens-Corning) had a $250,000 contract with the Air Force to develop radar-absorbent materials, and so the company president and vice president for research had secret clearances. Bissell was able to speak in general terms about his interest in a low-RCS aircraft. He obtained a list of personnel who would be involved in any work, and Jack Gibbs passed the list to the project security officer for expedited clearances [42].

TRACKING AND INTERCEPTION PROBABILITIES

On 2 January 1958, an updated assessment of Soviet interception capabilities was released by the CIA's Office of Scientific Intelligence (OSI) as an update to the October study. The only areas where the U-2 could fly without certain detection were central Siberia and east of Tashkent to China. Two new types of radar had been detected, one named STRIKEOUT and the other a modified TOKEN. It was now considered likely that an aircraft specially designed for operation at 70,000 ft would soon be available in limited numbers. SAMs were expected to become a serious threat in 1959 [43].

The Project Intelligence Section then went outside the Agency for confirmation of the findings. They spoke informally with air defense analysts and intelligence officers from the Army, Navy, Air Force, Air Technical Intelligence Center (ATIC), and the Strategic Air Command (SAC). They concluded that Scoville's new paper was more or less an average of the available opinions. There was apparently only loose agreement on SAM capabilities. A national intelligence estimate—from the previous March and generally considered to be in need of update—had described a SAM that could strike an aircraft at up to 60,000 ft within a 25-mile radius. Everyone expected that the overflights in the summer of 1956 had given the impetus to develop a new SAM with a maximum altitude of 70–90,000 ft, "depending upon who is doing the estimating." Probably no more than 20 of these were operational, but that 100 might be by the summer of 1958. Although the estimates from ATIC and OSI agreed, ATIC's was seen to be less the product of its analysts and more of a compromise to satisfy ATIC's leaders' desire not to underestimate the Soviet capability [44].

In the middle of the month, Bissell went to the White House along with five other senior advisors. Allen Dulles reviewed the AQUATONE operations and reported that some of the intelligence gathered was of "phenomenal" value. The president asked whether the Soviets had developed an interceptor that could operate at 70,000ft, and General Twining said that they were expected to do that by the end of the year. The group decided that the primary risk was of being discovered, and James Killian pointed out that the objective of the follow-on aircraft would be radar invisibility. The discussion moved on to a plan for a set of missions executed in rapid succession; the president felt that this would risk starting a war. Although it was not stated in as many words, the stakes for developing an undetectable aircraft had become even higher [45].

Chapter 4

THE FOLLOW-ON

At about the same time that Bissell was summing up the lessons learned from the first phase of RAINBOW, his admonition to Rodgers and Taylor to educate Kelly Johnson about radar began to bear fruit. However, even at this early stage they encountered the balancing act between low RCS and performance that has characterized the designs of all stealth aircraft. A conventional design was most likely to deliver the desired performance but could still be too easily detectable, whereas a design that emphasized stealth might not be able to deliver the needed range and payload.

B-2: A RESHAPED U-2

Johnson began to search for a way to design a relatively conventional aircraft that had some components to reduce RCS. His first concept was to build it of a thin Fiberglass shell and try to find a way to hide the engine and exhaust. He consulted Ed Lovick who explained that Fiberglass would be transparent at some frequencies, but would reflect at others. He also pointed out that the internal structure would have corner reflectors that would "sparkle."

To quantify how much energy would be reflected by the internal structure, the model group built a small-scale model with a Fiberglass skin and metal internal structure. They tried to make it as accurate as possible, including actually putting kerosene in the model's tanks. When it was measured in the anechoic chamber, the internal components indeed produced a large return. The fuel tank in particular looked like a large block of dielectric (Interview with Ed Lovick, Northridge, CA, 4 Feb. 2004).

Johnson then tried shaping the fuselage to reduce the return. Using his traditional carpenter's pencil, he sketched a fuselage that would reflect radar energy way from the transmitter (Fig. 15). The new design was an adaptation of the U-2 in which the bottom of the fuselage was blended into the bottom of the wings, the cylindrical fuselage was modified into one with relatively flat sides that were sloped inward to deflect radar energy upwards, and the leading edges of the wings were swept back. Only if the radar were aimed perpendicularly to one of the flat surfaces would it see a large return. He also

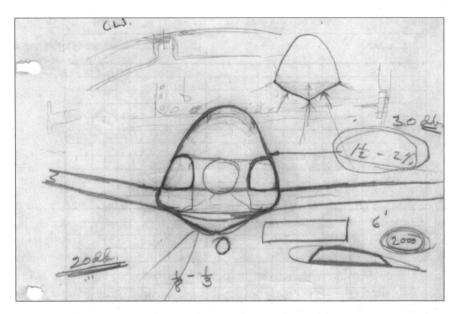

Fig. 15 Kelly Johnson's sketch of a reshaped U-2, with blended wings and fuselage and a wing box cross-section. (Courtesy of the family of Edward P. Baldwin.)

drew the trace of the inlet ducts, leading from the inlets on the sides of the fuselage to the single engine in the center of the fuselage; he hoped that the duct shape would hide the compressor face.

A wing cross section shows that Johnson was aware that some radar energy would penetrate the skin and reflect off internal components. The front and rear spars—"beams" in Lockheed terminology—were canted inward to reflect the energy upward.

Johnson made quick calculations of the radar reflections of Fiberglass, styrofoam, fuel, and an alloy called nickel zinc ferrite, to get a rough idea of the overall radar return of an aircraft with a Fiberglass skin and other components made of these materials. He also sketched out ways of fastening the skin to the frame. Then he turned the notes over to his "three-view man," Ed Baldwin, who produced a general arrangement (GA) drawing on 3 December 1957 (Fig. 16).

Ironically, this design was called the B-2. The name was a play on "U-2," although it anticipated the designation of the stealth bomber, developed by Northrop Grumman 25 years later. From above, the B-2 looked similar to the U-2. It was 65 ft long and had a wing span of 98 ft, making it significantly larger than the U-2A, which had a length of 49 ft 7 in. and a span of 80 ft. Its inlets were narrower than the semicircular inlets of the U-2. And, of course, the fuselage was no longer cylindrical, and there was no longer a right angle where it met the bottom surface of the wings.

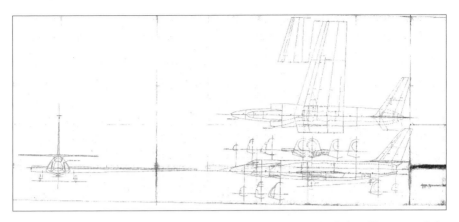

Fig. 16 B-2 design general arrangement drawing, by Ed Baldwin. (Courtesy of the family of Edward P. Baldwin.)

The B-2 apparently still had problems with the internal structure being visible to radar. A metal skin could avoid this by hiding the internal components, and so six days later Baldwin had produced a new version for comparison. The all-metal airplane, or AMA as it was called, substituted metal for the B-2's Fiberglass skin (see Fig. 17). The objectives apparently were to improve reflection of radar energy at "innocent angles"—away from the radar receiver—as well as to hide the internal components of the aircraft, like the engine. However, the design still had what Baldwin called "bad corners" in the area of the horizontal stabilizer, which would "shine like a sore thumb" when illuminated by radar.

As each design took shape, an RCS model would be produced and handed over to Mel George's team for measurements. Jim Herron eventually filled two filing cabinets with polar plots of the returns of various models.

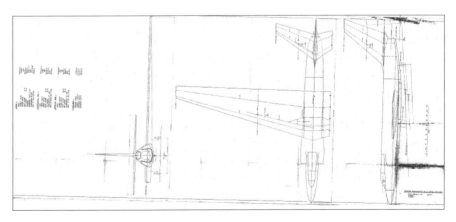

Fig. 17 All-Metal Airplane general arrangement drawing, by Ed Baldwin. (Courtesy of the family of Edward P. Baldwin.)

GA #2

Through December and January, the team worked on design variations that would solve not just the horizontal stabilizer problem, but other sources of reflections. The compressor blades would "shine brightly," and the exhaust would also reflect radar well, in addition to being a source of infrared (heat) emissions. And if the leading or trailing edge of the wing were straight, then it would produce a sharp reflection when the radar beam was at a right angle to the edge.

Baldwin explored variations of elliptical and parabolic wings, attempting to work out the shape of a wing that—while reflecting the minimum radar energy—would still be able to support the aircraft in stable flight from ground level up to the operational altitude (see Fig. 18).

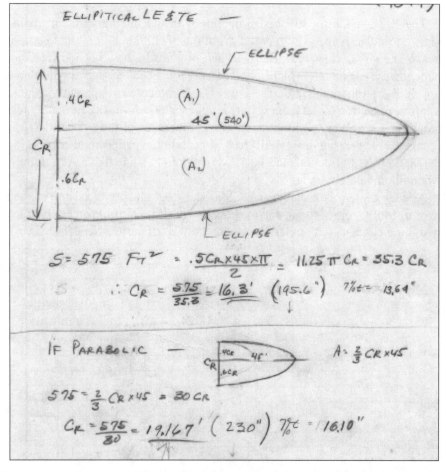

Fig. 18 Design of a wing with elliptical leading and trailing edges, by Ed Baldwin. (Courtesy of the family of Edward P. Baldwin.)

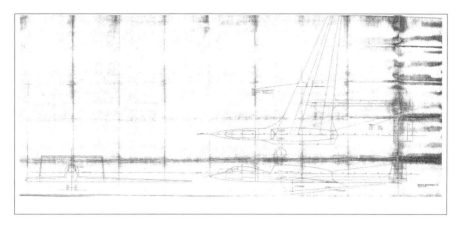

Fig. 19 General Arrangement #2 general arrangement drawing, by Ed Baldwin. (Courtesy of the family of Edward P. Baldwin.)

By 5 February 1958, he had worked out a design known simply as "General Arrangement #2" (Fig. 19) whereas the B-2 and the AMA were obviously descendants of the U-2, GA #2 was a totally new design. It had trapezoidal wings, twin booms, vertical tails that tipped inwards, a horizontal tail atop the vertical tails (similar to that of the OV-10 Bronco), and a single engine with its inlet and exhaust on top of the fuselage. The technique of placing the inlet and/or exhaust above the wing and set back from the edge in order to reduce its exposure to radar was used in numerous later stealthy designs that have actually flown, including HAVE BLUE and the F-117, TACIT BLUE, and the B-2 Spirit stealth bomber.

GA #2 appears to have been presented to the CIA as "GUSTO Model I," a more-or-less conventional design that presented the least risk in terms of performance.

As is common when designers must work through numerous concepts, the structural work for GA #2 was never completed. The main objective was to determine the design's feasibility, and so work focused on the novel parts of the design where unexpected problems might lurk. In particular, the booms and elevated horizontal tail were uncommon and received lots of attention to ensure that they could be built sufficiently strongly, and one boom was actually built. On the other hand, the wing structure was conventional so that its design did not require much detail; an aerodynamic analysis to find the wing's mean aerodynamic chord (MAC) was sufficient. Because the 25% line of the MAC must be located at the complete aircraft's center of gravity, the calculation was needed in order to determine how far forward or aft the wing assembly would be placed. The aerodynamic analysis showed that GA #2 would probably fly no higher than the U-2.

The skin of GA #2 was a complex structure. The center was a 7/10-in.-thick sheet of honeycomb. On each side of the honeycomb was $\frac{1}{8}$ in. of

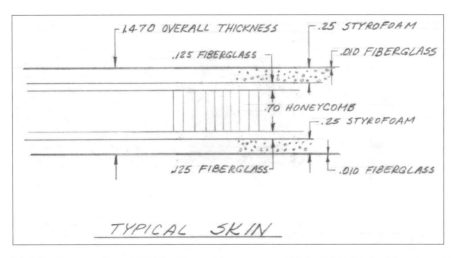

Fig. 20 Cross-section of skin for General Arrangement #2, by Ed Baldwin. (Courtesy of the family of Edward P. Baldwin.)

Fiberglass, then $\frac{1}{4}$ in. of styrofoam, and finally a 1/100-in. layer of Fiberglass, for an overall thickness of 1.47 in. (see Fig. 20).

Project GUSTO

During January 1958, the Agency assigned the cryptonym GUSTO for the RAINBOW Phase II work. Starting a new project was always an opportunity to tighten security, in this case by excluding people who had known about the RAINBOW antiradar work, but who had no need to know about the follow-on aircraft. The new name also made it difficult for outsiders to connect the old and new projects. On 28 January, Agency (CIA) personnel came to Burbank to give the initial security briefings for GUSTO. The Agency went so far as to assign a new internal mailbox for GUSTO correspondence [46].

On 30 January, Johnson wrote a letter to Bissell proposing a work statement for Project GUSTO. There were four basic tasks:

> 1) Carry on preliminary design of the bird shown to you during our recent visit, including studies of equipment arrangement, structure, weight, balance, etc.
> 2) Undertake early wind-tunnel tests in both a low-speed and high-speed wind tunnel.
> 3) Carry on such tests as we are able to make on the effect of various materials and shapes of the bird with regard to their reflective characteristics.
> 4) Investigate various plastic structures for use in certain areas of the bird. This includes structural testing, availability, and desirability of subcontracting certain construction.

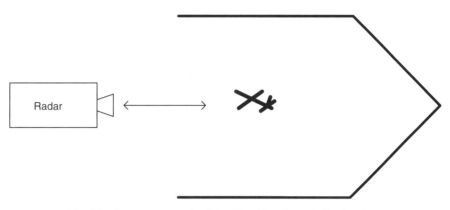

Fig. 21 Arrangement of original Lockheed anechoic chamber.

Johnson expected that by June he would have completed low-speed tunnel tests and be well into construction of a model for the high-speed tests. The statement of work was approved on 11 February [47].

In early 1958, Lockheed built its first anechoic chamber for testing the models (see Fig. 21). Located in Building 82, it was 12 ft wide, 12 ft high, and 30 ft long. Compared with the large chamber that would be built in 1964, the original one was primitive. As SEI had found, high-quality radar absorbent material was not available in large quantities. The chamber was lined with flat sheets of material provided by Emerson and Cuming, who were producing the material applied to the T-33 for the PASSPORT VISA tests.

The RAM had front and back layers and in between them a dielectric material with ellipsoid-shaped voids. The front layer tended to reflect some energy, which interfered with the measurements from the model and reduced the precision with which its RCS could be determined (see Fig. 22).

Models were suspended on strings from the ceiling of the chamber and illuminated by the radar (see Fig. 21). Stray reflections would strike the walls and be mostly absorbed. All of the aircraft designs developed in the course of work on the U-2 follow-on were tested in this chamber and were never sent outside Lockheed for additional testing (Interview with Ed Lovick, Northridge, CA, 8 March 2008).

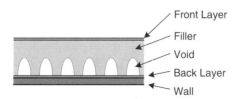

Fig. 22 Cross section of radar absorbent material from original anechoic chamber.

More RAINBOW Flights

As GUSTO ramped up, work continued to perfect RCS measurement techniques not only to improve the protection on the Dirty Birds, but also to gather data for designing the follow-on. A mid-January memo probably from SEI described making indoor measurements on "shaded" ellipses and other shapes at S-, C-, and X-band, with K-band soon to follow. A new technique to soften the wing edge that used a uniform resistance sheet cut in a sharks tooth pattern had been discovered. This was expected to be easier to apply to aircraft than previous materials.

Some progress had been made on electrically shading Fiberglass structures. Using a $\frac{1}{8}$-in.-thick slab of Fiberglass laminate as the simulated structural member, a quarter-wavelength layer of dielectric was applied to both sides, which is a technique used to coat optical lenses. There were experiments with varying numbers of layers of dielectric, as well as with a single thick layer with a continuously varying dielectric constant. The next step would be to load a lightweight foam with aluminum flakes. This work was expected to be so successful that the airframe would no longer be the major source of RCS; "the major contribution to the cross section should now be from the engine and equipment bay which can be seen through the transparent fuselage and to some extent past the protective discs especially at higher frequencies" [48].

The systems for measuring actual aircraft on a pole had been greatly improved. The data-collection system had been modified to record the range, azimuth, and elevation of the target along with the magnitude of the return. This would make data reduction much simpler and less error prone. The pole system for ground measurements was showing "a fantastic amount of detail all of which is reproducible to the minutest detail on successive runs." The return from the target was far above the background clutter, and the pole had been shielded to keep its return well below that of the target.

It was recommended that one system be disposed of, as it was of no further use. The author of the report commented that

> In an operation such as this in which we flagrantly violate established government procedures in order to get a job done rapidly, it behooves us to keep our inventory to a minimum both in order to keep the operation small and flexible and minimize the number of years we may eventually have to spend in jail.

Finally, the report laid out a series of tests using actual U-2s. Article 359 would be measured at different elevation angles, frequencies, and radar beam polarizations, and the system would be checked for reflections between the aircraft and the ground under it. Article 378 would then be used for measurement on the tail pipe and other parts of the aircraft for which no protective measures had been devised. Article 367 would then be tested. Tests on a human were being planned, so that they would know whether the pilot could be left unshielded in the follow-on aircraft [48].

During late February and early March, four of the U-2s (344, 349, 355, and 367) were cycled through Burbank to have their Wallpaper coating repaired. To be sure that the patches were responding properly, each repair job was followed with a series of test flights before the aircraft was ferried to its operational detachment. The ferry flights were to be completed by 24 March [49].

On 13 March, Leo Geary reported that U.S. radars had tracked article 355 on its outbound ferry flight. The readings were very accurate in position and altitude. Trapeze had not been installed at the time of the flight, and that was assumed to be the reason that tracking was so easy [50].

In April 1958, SEI was planning to begin testing large-scale models at 600 MHz. There was a question of whether to do this on the existing range at Indian Springs Air Force Base or to move to the Ranch, where the security was better. At that time, Indian Springs had a single pole 5000 ft from the antenna array, and there was a difference of opinion as to whether there would be too much interference from ground clutter for measurements at 600 MHz. If it could not be used, then a new pole would have to be added at 1500 ft from the antennas and at a right angle from the line to the existing pole. Frank Rodgers dispatched one of the engineers to Indian Springs to make definitive measurements.

Gene Kiefer summarized this in a memo to Bissell and also indicated the need to procure a second "knuckle," which was the name for the connection between the model and the pole. When a model was mounted, it was first attached to the knuckle, and the knuckle was then attached to the pole. Having a second knuckle would speed up the process of swapping models on the pole because "... considerable time is lost in changing the knuckle from one type of model to another." The second knuckle would be needed anyway if a second pole were installed. Bissell approved the decision to stay at Indian Springs through the summer and to procure the second knuckle [51].

GA #3: The Batplane

The Lockheed team next took up the higher risk designs, in which low RCS was the top priority. They would first shape the airplane to minimize RCS and then try to find a way to make it fly with the performance needed to meet the mission requirements. These designs became known to the Agency as GUSTO Model 2.

Work continued into April 1958 on these designs in an attempt to remove all straight lines; these were referred to by Baldwin as General Arrangement #3. Ray Kirkham sketched a conceptual design known as the "Batplane," which had all curved leading and trailing edges, a blended

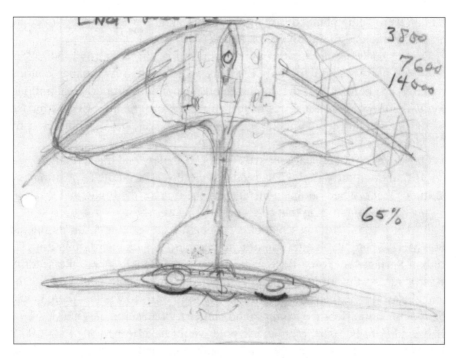

Fig. 23 Sketch for the "Batplane," by Ray Kirkham. (Courtesy of the family of Edward P. Baldwin.)

wing and body, and twin engines buried in the wings (Fig. 23). He also sketched a flying wing, a design with no empennage. The aircraft would have used a pair of either J79 or J52 engines.

Baldwin worked on a "scimitar" wing—which had no straight edges—for GA #3 (see Fig. 24). Analyzing the aerodynamic properties of such a complex shape, using nothing more that a slide rule or mechanical calculator, took a long time because there were quadratic equations to solve all along the length of the wing.

GUSTO 2: FLYING SAUCER GROWS WINGS

The next design began with the flying saucer. The cockpit was placed in the center of the saucer—actually more of a triangle with rounded corners—between two engines blended into the upper surface. The edges of the saucer were then notched to accept wedges of graded dielectric material, an invention of Ed Lovick.

The concept of the triangular material came from the pyramids of radar-absorbent material used in anechoic chambers and was, according to Lovick, "no great secret." When the radar energy entered the material, it would be reflected inside the pyramid, generating electrical currents that turned the radio frequency energy into heat. When the idea was applied to the chines

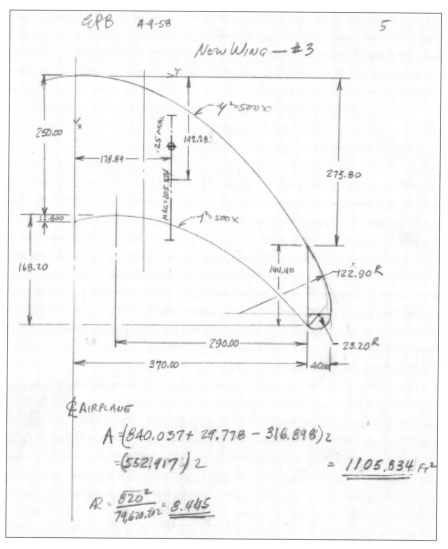

Fig. 24 Scimitar wing design for GA #3, by Ed Baldwin. (Courtesy of the family of Edward P. Baldwin.)

of Lockheed's A-12 design, the space inside the chines was needed for plumbing and equipment, and so wedges were tried initially but eventually were replaced with sheets of graphite-impregnated asbestos honeycomb.

This was the technique called "softening" the edge of the wing. The purpose was to avoid having an abrupt change in resistivity at the edge or farther back in the wing; abrupt changes cause reflections. Because the impedance of free space is 377Ω, that should be the impedance of the wide side of the triangles at the wing edge, where the radar energy first meets the aircraft. The resistivity should decrease to zero at the back, matching that of the adjacent metal structure. Internal reflections along the sides of the triangles would keep the energy inside the material, and it would be dissipated as heat.

Because the saucer itself did not have enough lift, the next step was to add all-plastic wings, resulting in the GUSTO 2 design (see Fig. 25). However, RCS measurements of this design determined that using only plastic did not guarantee radar invisibility. Nearly 20 years later, Kelly Johnson summarized the problem in a paper for the Radar Camouflage Symposium, a classified conference on stealth aircraft: "It was found, for instance, that when the plastic parts, such as those that might be designed in wing beams, or in heavy structural rings exceeded $\frac{1}{4}$ or $\frac{1}{2}$ in. in thickness, these members might just as well be metal" [52].

There were a number of other lessons from the studies of plastic components. One was that at some frequencies a plastic skin was effectively transparent to radar, and that internal components like the engine provided hundreds of corner reflectors, which produced a large return. Although metal components might seem an obvious source of reflections, fuel was not. One of the most unexpected findings was that a vibrating fuel tank would have standing waves in the surface of the fuel, which would produce a much larger

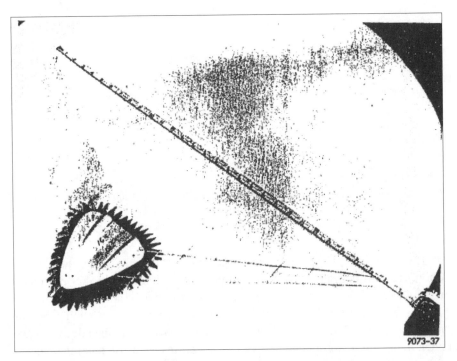

Fig. 25 GUSTO 2 model with plastic wings. (Courtesy of Lockheed Martin and Jeffrey Richelson.)

return than a motionless tank with no waves. However, it was not possible to prevent vibrations in the fuel, so that any aircraft with transparent skin had yet one more source of reflections.

The final design, GUSTO 2A, used the wedge-shaped inserts along the leading and trailing edges of its wings to prevent radar echoes from its internal structure (Fig. 26). It had a single inlet in the nose of the saucer and vertical stabilizers with rudders at the ends of the wings (see Fig. 27).

For comparison, an all-metal version, GUSTO 2S, was tested. Radar tests confirmed that the treatment on GUSTO 2A gave it a significantly lower RCS than the all-metal version [53].

GUSTO 2A could not meet the mission requirements. Although it could achieve the desired 2000 mile radius, it could only reach an altitude of 75,000 ft, where its RCS would still be too high. Higher altitude would have helped, but that was not achievable with the modest coefficient of lift of a flying wing. An aircraft can achieve higher altitudes by using a wing with more camber, which has a higher coefficient of lift. The disadvantage is that the airflow over the wing will tend to make it pitch downward. A conventional design compensates for this by having a small horizontal stabilizer far aft of the wing to apply downward pressure and thus trim out the downward-pitching moment (see Fig. 28).

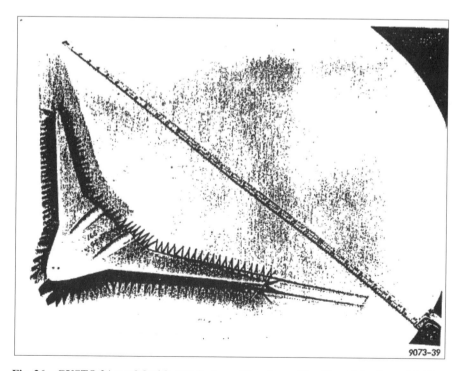

Fig. 26 GUSTO 2A model with metal wings. (Courtesy of Lockheed Martin and Jeffrey Richelson.)

Fig. 27 Dan Zuck (left) and L. D. MacDonald (right) with the GUSTO 2A model. (Courtesy of Lockheed Martin.)

Fig. 28 Conventionally cambered wing and tail.

A flying wing, on the other hand, has no separate tail and must compensate for the pitch-down moment by having the trailing edge of the wing turn upwards (called reflex camber) to deflect the air flow upwards and push the trailing edge down (see Fig. 29). Unfortunately, this results in a wing with a lower coefficient of lift, which limits the maximum altitude. If the wing is swept, the

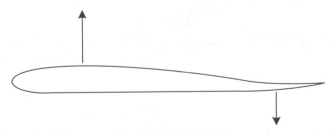

Fig. 29 Reflex cambered wing.

reflex camber can be used on only the outer parts of the wing, which are far enough behind the center of gravity to have a larger moment arm and keep the aircraft trimmed. This means that the inner sections of the "twisted" wing will have a better coefficient of lift than the outer. However, in the case of GUSTO 2A, the final wing design still could not lift the weight—including structure, radar-absorbent materials, engines, fuel, pilot, and cameras—to an altitude where its RCS would be low enough to avoid detection.

PLASTICS

While these designs were being explored, both the CIA and Lockheed were exploring sources of composite materials. On 10 April 1958, Jack Gibbs visited Owens-Corning to learn whether they could provide structural materials. One of the possible materials discussed for building a plastic airframe was flake glass. It turned out that they did not yet have enough experience with the material to understand its elasticity and other physical characteristics to give any confidence that it could be used for load-bearing structures in an airframe. As glass producers, Owens-Corning had never investigated adhesives, which would have been needed to assemble plastic components. Moreover, they had no experience in analyzing or building load-bearing structures, like airplanes, and no experience with fittings to join Fiberglass and metal components, as Gibbs and the team envisioned. Nevertheless, the company could see a potential market and expressed interest in getting a development contract to nail down the basic properties of Fiberglass that would be needed to use it as a structural material. They invited Gibbs to visit their development laboratory in Ashland, Rhode Island, to learn more [54].

Five days after the Owens-Corning (O-C) trip, Gibbs was at the Aircraft and Materials Labs at Wright Field, to learn what work they were doing with composites and what they had learned. He told them that the Air Force was interested in getting the aircraft industry to use plastics, such as Fiberglass, in aircraft design and wanted to pass on what they knew to Richard Horner, assistant secretary of the Air Force for research and development.

In discussing Owens-Corning's work on flake glass, Gibbs was told that although individual glass fibers or flakes had the same strength as aluminum, when they were combined into a larger structure by embedding them in resin or weaving them into a cloth and impregnating the cloth with resin, the strength of the finished product was only about one-third of aluminum's. Long term, they thought the strength of the final material could be doubled, which would still be less than that of aluminum.

The Materials Lab was funding work on ways to improve this, including a contract with Owens-Corning to investigate beryllium-oxide fibers, which had a basic strength twice that of flake glass. However, they felt that O-C was for unknown reasons dragging their feet, and Gibbs concluded that it would be at least a year before there would be usable materials. The lab personnel felt that flake glass was still very experimental, and not enough work had been done to produce samples that could be evaluated. They thought that the $15,000 three-month study which O-C had proposed to Gibbs was much too small to produce useful results.

Besides the Owens-Corning contract, Wright Field also had contracts with two other companies to investigate adhesives for plastics and ceramics and to investigate high-strength fibers. When Gibbs asked the Materials Lab about who could design with Fiberglass, they were unanimous that the aircraft industry lacked the skills, and that "the aircraft industry are sheet metal workers and just don't know how to get the most out of plastics designing" [55]. Instead they recommended two companies for Fiberglass design, one of which was Goodyear.

In his report to Bissell, Gibbs recommended a way to move the project forward using Fiberglass structures. Lockheed would finish the Gusto design in all metal and perform a stress and RCS analysis. Then the drawings would be marked to show which parts would be Fiberglass; these drawings would be given to the Fiberglass contractor to design the Fiberglass parts and fittings to connect metal and Fiberglass. Then a stress analysis would be done for the composite design and the design presented to Lockheed for review. If there was agreement, then a prototype would be built, with Lockheed doing the basic metal airplane and Goodyear (or the alternate) building the plastic parts. They and Lockheed would then work together on final assembly. If the finished airplane could pass static testing to 50% above the designed limit load (like the U-2), then the airplane would be flown. Successive airplanes would also be static tested; this would give the most confidence in this radically new design approach [55].

Whether this plan was ever presented to Lockheed is not known.

At Lockheed, Mel George was also in contact with Corning on composite materials and, like Gibbs, decided that it was unlikely that they would be able to produce high-strength material usable for aircraft structures in the near future. In particular, he concluded that the strength of the materials would not be increased by more than 30 to 35%. Hearing his report, Johnson decided that a composite structure of the required strength would actually be heavier than an all-metal structure and could not meet the altitude and range requirements.

The prospects for a stealthy, high-altitude, subsonic aircraft were not looking good.

Chapter 5

HIGH SPEED

Even before the U-2's operational missions had begun, Johnson was looking ahead to the next leap in aircraft performance. For two-and-a-half years—in parallel with the deployment of the U-2, with the antiradar work of Project RAINBOW, and with the subsonic GUSTO designs—Lockheed was funded by the Air Force in a series of projects to develop a Mach 2.5 high-altitude reconnaissance aircraft. Although only one of these designs made it even as far as a mock-up, the work set the stage for the next major step in project GUSTO, a Mach 3 design. (Primary documents on the hydrogen-powered aircraft described here are mostly unavailable. This chapter draws on John Sloop's *Liquid Hydrogen as a Propulsion Fuel, 1945–1959*, Jay Miller's *Lockheed Martin's Skunk Works*, and Dick Mulready's *Advanced Engine Development at Pratt & Whitney*.)

REX AND THE CL-325

In the mid-1950s, liquid hydrogen appeared to numerous agencies and companies to be the aircraft fuel of the future. One of the key technologies, a liquid-hydrogen-fueled aircraft engine had been invented by Randolph Rae. Rae was a former employee of the British Admiralty who had come to the United States in 1948 and gone to work in the Applied Physics Laboratory of Johns Hopkins University (APL/JHU). He became interested in the problem of powering a subsonic aircraft at very high altitudes and saw a possibility in using a hydrogen-powered rocket engine to produce a stream of gas that would spin a turbine, with the turbine driving a large propeller. In need of a corporate sponsor, he joined Thomas Summers, of Summers Gyroscope. Summers also hired Homer J. Wood, a consultant formerly with Garrett Corporation.

Rae and Wood developed what they called the Rex engine and in March 1954 presented the proposal, "REX-I, A New Aircraft System," first to ARDC and then to the Wright Air Development Center (WADC). Powered by liquid hydrogen and liquid oxygen, REX-I could fly at 85,000 ft with a range of over 6000 miles. By the spring of 1955, Garrett Corporation had bought Summers Gyroscope, and the Air Force was giving other companies contracts to investigate using hydrogen fuel in turbojet engines. The high-altitude flight regime was of interest because the Air Force was eager to take back the lead in aircraft technology from the CIA, sponsors of the U-2.

Rae and Garrett doomed their proposal in contract negotiations by insisting that Garrett would be the contractor for the entire airplane, which was contrary to the Air Force's usual process of having the airframe manufacturer as the prime contractor and the engine manufacturer as a subcontractor. Moreover, the ARDC powerplant laboratory was negotiating for the Air Force and wanted to see a series of steps proving the practicality of the hydrogen engine before committing to an airplane using the engine. As the negotiations proceeded, Rae's successive proposals became very complex. The engines evolved into large turbojets using external air, rather than liquid oxygen, and the airplane evolved into a supersonic one with shorter range.

In October 1955, Rae and Garrett finally signed a contract with the Air Force and subcontracted with Lockheed to study the design of a supersonic aircraft using 48-in.-diam Rex engines. Despite the fact that Garrett had never built a turbine engine much larger than 8 in. in diameter, they did not foresee problems. Johnson's team chose a design point of Mach 2.25 and soon decided that a 60-in. engine would be needed. Garrett agreed to the new requirement, and the Rex III engine was the result. It incorporated a new feature for turbojets: a heat exchanger using exhaust gases.

On 9 November, during the early stages of the study for Garrett, Johnson visited the Pentagon to discuss building U-2s for the Air Force. He had earlier made a verbal proposal to the Assistant Secretary of Defense, Trevor Gardner, and was now coming for a face-to-face discussion. The meeting covered various U-2 configurations, including an interceptor version (rejected until a U-2 could be tested intercepting a B-52) and a reconnaissance version identical to the CIA's, which he referred to as the L-182. (The use of "L-182" to refer to the U-2 was a subterfuge. The actual L-182 design was a 1948 paper design to update the avionics and armament of the P2V-4.) But Johnson also undertook to study a hydrogen-powered U-2.

After less than three weeks, the results of the study were "quite adverse." The U-2 could only carry 340 gal of hydrogen, the equivalent in heat content of a mere 70 gal of JP-6, which would be "insufficient for safe flight." Drop tanks could add a small amount, but not enough to make a significant difference, and "the volume problem looks almost insuperable." Johnson rescinded the proposal, unwilling to spend more effort on it [56].

By the end of January 1956, the Lockheed study for Rae was complete. The CL-325-1 was 155 ft long and had a straight wing with a span of 81 ft. Like the F-104, the CL-325 had a T tail, and the wing had anhedral (negative

HIGH SPEED

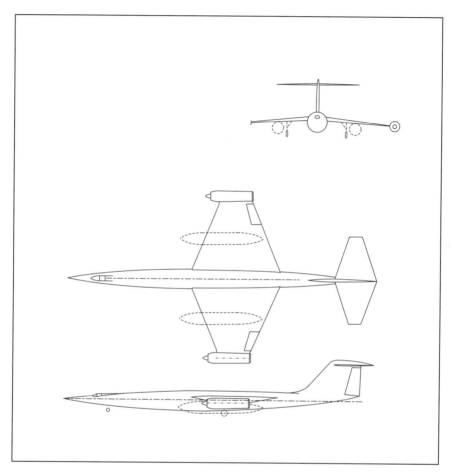

Fig. 30 CL-325-2 design. (Drawn by author from Lockheed Aircraft Corporation drawing.)

dihedral) to reduce the Dutch roll problem that would have resulted. The engines were placed at the ends of the wings to reduce the weight concentrated in the fuselage and the resulting bending moment on the wings. The fuselage was thin and contained a single hydrogen tank. At takeoff it weighed 45,000 lb. Its range was only about 3100 miles. A second version, the CL-325-2, placed some of the liquid hydrogen in droppable wing tanks, which reduced the size and weight of the aircraft by about 15% (Fig. 30).

Garrett presented the complete study at Wright Field on 15 February 1956. There were few questions, as the Air Force team soon realized that Rae's original idea of a small, simple engine and airplane had evolved into a large, complex turbojet powering an aircraft longer and faster than a B-58. They decided that Garrett lacked the facilities, the tools, and the experience to take on the project. Meanwhile, experienced engine manufacturers were also proposing hydrogen-fueled turbojets. Garrett continued for another year on two more studies, but in October 1957 they were directed to stop work.

SUNTAN

Having completed his contract with Garrett, Johnson apparently decided that a liquid-hydrogen-powered aircraft was practical, but would require a different engine contractor. Although all of his motivations are not documented, Johnson was too good a politician not to know that the CIA's success with the U-2 had left the Air Force wanting to retake the lead in aircraft performance. In January 1956, he visited Air Force Lt. General Donald Putt, the deputy chief of staff for development. He proposed a Mach 2.5, 100,000-ft-altitude liquid-hydrogen-powered reconnaissance aircraft. He promised to deliver a first flight within 18 months of approval.

On 18 January, Putt convened a meeting to evaluate the Lockheed proposal. He invited various Air Force commanders, including Lt. Gen. Clarence Irvine, the head for materiel; Lt. Gen Thomas Power, head of ARDC; and Col. Norman Appold, head of WADC's powerplant laboratory. The committee's evaluation was positive, but Putt wanted a six-month evaluation period to verify the feasibility of the expected performance. Appold was put in charge of the ARDC evaluation team. His first action was to contact both General Electric and Pratt and Whitney (P&W). Both companies submitted proposals, and he selected P&W; the selection was approved on 20 February.

The overall program was code-named SUNTAN and was one of the most secret programs the Air Force has conducted. Only 25 people were authorized access to all of its aspects. Expenditures were carefully hidden; the amount of money spent was at least $100 million and might have been more than twice that.

By the spring, the Air Force had signed contracts with both Pratt and Whitney and Lockheed. Lockheed was to build two prototypes, and six production aircraft were soon added to that. Work was actually underway before the contract was signed. Johnson assigned Henry Combs to be the project engineer on the new aircraft, which was named the CL-400 (Telephone interview with Henry G. Combs, 3 June 2001).

The resulting CL-400-10 was a slightly larger version of the CL-325 with a ventral fin added to improve lateral stability, which allowed the now-unnecessary wing anhedral to be removed (see Fig. 31). In place of the Rex III engine at the end of each wing was a Pratt and Whitney Model 304-2. Unfortunately, the percentage of total weight available for fuel was reduced by 3%, and the radius dropped from 1553 to 1130 n miles, a loss of 27%.

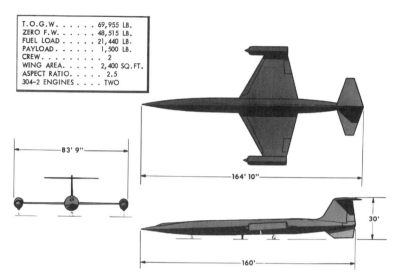

Fig. 31   CL-400-10 design for SUNTAN. (Courtesy of the Jay Miller Collection and the Aerospace Engineering Center.)

Ultimately, SUNTAN was doomed by a cascading series of problems, all related to the production and storage of liquid hydrogen. The first problem was storage in the aircraft. To prevent cryogenic liquids from boiling off, they must be stored in dewars, insulated containers. To reduce the leakage of heat into a dewar, the ratio of the surface area to the volume must be minimized. A sphere is the best possible shape, but a cylindrical tank with hemispherical ends—which would fit in an aircraft fuselage—is a reasonable compromise. A wide shallow tank—the ideal shape to fit in a wing—is not. Johnson explained it by saying, "we have crammed the maximum amount of hydrogen in the fuselage that it can hold. You do not carry hydrogen in the flat surfaces of the wing" (Sloop, J. L., *Hydrogen as a Propulsion Fuel, 1945–1959*, SP-4404, NASA History Series, National Aeronautics and Space Administration, Washington, DC, 1978, pg. 164). (In principle, many small cylindrical dewars could be fit into the wings, but the increased capacity is not necessarily worth the increased weight.)

An airplane's range is determined by the specific fuel consumption (SFC) of its engines, the lift-to-drag ratio (L/D) of its wings, and the percentage of its weight used for fuel (the fuel fraction). Range can be increased by decreasing the SFC, by increasing the L/D, or by increasing the fuel fraction. The SFC of the Pratt and Whitney engines was not bad. However, the L/D of the wing was not good enough. A larger wing would have given a better L/D, but at the cost of an increase in weight. In an airplane burning hydrocarbon fuels, a larger wing gives more space to store fuel, so that the fuel fraction will not suffer. However, SUNTAN's wings could not contain dewars, and so the increased L/D of the larger wings would come at the cost of a reduced fuel fraction, causing a decrease in range.

The only way that a CL-400 could be flown over the Soviet Union would be by basing it in a foreign country. Aggravating the usual problems of politics and security would be providing the liquid hydrogen to fuel the aircraft. Either it would have to be shipped from the United States, or it would have to be made on site, consuming enormous amounts of natural gas and electricity. Neither alternative was practical. Ben Rich summarized the situation by asking, "How do you justify hauling enough LH_2 around the world to exploit a short-ranged airplane?" (Sloop, J. L., *Hydrogen as a Propulsion Fuel, 1945–1959*, SP-4404, NASA History Series, National Aeronautics and Space Administration, Washington, DC, 1978, pg. 164).

Kelly Johnson recognized the problem, and in mid-1957 recommended to James Duncan, secretary of the Air Force, that the project be cancelled. ARDC disagreed with Lockheed's range estimates and declared the design capable of a range of 3500 n miles. In late 1957, after much argument, the funding was cut, although Pratt and Whitney was given $18.7 million to continue engine development. Lockheed discontinued work on a mock-up, but was paid to continue fuel system tests. They returned $3 million from money previously received.

While Lockheed reduced work on the project, Pratt and Whitney continued at full speed with development of what had become the Model 304 engine

Fig. 32   Model 304 liquid hydrogen engine at Pratt & Whitney Florida Research and Development Center. (Courtesy of Roadrunners Internationale.)

(see Fig. 32). An Air Force C-124 carried the first engine from the company's development facility in East Hartford, Connecticut, to their Florida Research and Development Center on 19 August 1957, and the first tests began on 11 September. The roar of the engine resulted in its nickname, "The Swamp Monster."

More Studies

The Air Force requested further airframe studies from not just Lockheed, but also North American, Boeing, and Convair. Lockheed studied 14 different aircraft for both reconnaissance and bombing missions, at speeds up to Mach 4, and compared hydrogen and hydrocarbon fuel. Most had trapezoidal wings like the CL-325 and CL-400-10, but with the engines mounted in either underwing nacelles or under the fuselage. Eventually, the CL-400 designs reached an enormous size, while still carrying only two crew members and 1500 lb of payload. The CL-400-13 would have used two proposed STR-12 hydrogen engines (scaled-up Model 304s), had a takeoff gross weight of 376,000 lb, and had 6500 ft^2 of wing area (Fig. 33). The delta wings were far to the rear, and a pair of canards were behind the cockpit.

At a takeoff weight of 158,620 lb, the CL-400-15JP was a much smaller design. It used two J58 engines burning hydrocarbon fuel rather than hydrogen and fed air through a single large half-cone inlet on the bottom of the fuselage.

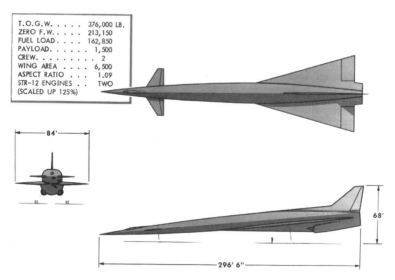

Fig. 33   CL-400-13 was among the largest of Lockheed's SUNTAN follow-on designs. (Courtesy of the Jay Miller collection and the Aerospace Engineering Center.)

It was 144 ft 6 in. long and had a wing span of 56 ft 6 in. and a wing area of 1800 ft^2. It also used canards.

Comparing the hydrogen- and hydrocarbon-powered designs, Lockheed found that for a given range, a hydrogen-powered aircraft would be larger while weighing less at takeoff because of the low density of hydrogen compared to hydrocarbon fuel. For a fixed cruise speed, the hydrogen-powered aircraft would have an altitude that was higher by 10,000 to 20,000 ft. Overall, Johnson concluded, using hydrocarbon fuel resulted in a smaller and less expensive design, thanks to the higher energy density of the fuel. Moreover, the speed could be increased to Mach 3, at the cost of 10% of cruise altitude. Although the heat was significantly more than at Mach 2.5 and aluminum could not be used, titanium appeared to be a viable alternative.

The designs from all four contractors were evaluated by the Air Council on 12 June 1958. Boeing had produced the most promising design, a 200-ft-long aircraft with a delta wing having a span of 200 ft, powered by four engines burning liquid hydrogen. The radius was almost 2300 n miles, twice that of the CL-400. However, given the political risks, there was uncertainty over whether the President would allow its use, and the Air Force might just be building a museum piece. Even worse, another Air Force project had already been given higher priority, and it needed more money. The Air Council decided to terminate the SUNTAN funding, except for engine work. Final runs of the Model 304 engine were made in late September 1958. Five engines had been built, and the total run time on hydrogen added up to only about 25 hours.

SUNTAN was finally over, although it led to two significant developments. Pratt and Whitney's two-and-a-half years of experience with liquid hydrogen made possible the RL-10 rocket engine. Installed in a variety of missiles, the RL-10 eventually launched scores of satellites and space probes and has continued to evolve into the 21st century.

The other development was that Lockheed now had an improved understanding of high-speed flight as well as a confirmation that hydrocarbon fuels were the best choice for that flight regime. Johnson was now ready to propose a major change in direction for Project GUSTO.

GUSTO Supersonic Designs

In April 1958—probably as a hedge against the problems he was finding with the subsonic GUSTO designs, and while his team was still working on designs for the final round of liquid-hydrogen aircraft proposals for the Air Force—Kelly Johnson began adapting the CL-400-15JP for higher speed. Although the new hydrocarbon-fuel-powered aircraft would not be able to reach a 100,000-ft altitude, it would be able to reach a speed of Mach 3.

On Monday, 21 April he began sketching his first Mach 3 design. For a name, he chose "U-3," presumably because it was a successor to the U-2. Over the course of three days, in between his other work, he laid out the basic requirements, estimated the size of the aircraft, and did a rough sketch:

High-altitude cruise:	90,000 ft
Design cruise Mach No.:	3.0
Engines:	Two
Crew:	Basic – one (two in future)
L/D reqd.:	7 to 8
Range:	2000 n mile radius
Payload:	500#

Johnson, C. L., Archangel project design notebook, Lockheed ADP, Burbank, CA, entry for 21 Apr. 1958. pg. 1.

Johnson compared the expected performance of J58 and J93 engines at the cruise altitude and speed:

	J-58	J-93
Thrust	4000 lb	2800 lb
SFC	2.4	2.8
Weight including afterburner	5900 lb	4370 lb
Thrust/weight	0.677	0.64

Johnson, C. L., Archangel project design notebook, Lockheed ADP, Burbank, CA, entry for 22 Apr. 1958. pg. 2.

Choosing the J58 because of the higher thrust-to-weight ratio and lower specific fuel consumption, he then began computing weights. With an assumed lift-to-drag ratio of 7.5 (in the middle of the expected range) and 8000 lb thrust from two J58s, the maximum weight at altitude was 7.5 × 8000, for a weight of 60,000 lb at the start of cruise. He then solved the equation for the amount of fuel needed to make the mission range (Johnson, C. L., Archangel project design notebook, Lockheed ADP, Burbank, CA, entry for 22 Apr. 1958. pg. 2):

$$\frac{\text{range}}{\text{speed}} \times \text{thrust} \times \text{SFC} = \frac{4000}{1770} \times 8000 \times 2.4 = 43{,}400 \text{ lb}$$

Subtracting the fuel and engine weight from the 60,000 lb target gave only 5000 lb, about 80% of the weight of the U-2's airframe, an unlikely target.

To see whether a better wing would help, he tried an L/D of eight and got an airframe weight of 9000 lb out of a total weight of 64,000 lb. With a coefficient of lift of 0.2 and dynamic pressure of 230 psf, a wing should have provided 46 psf of lift. Dividing that into the total weight, this would have mean that the airplane needed 1400 ft² of wing. At best the wing would have weighed 2800 lb and would have had a span of about 37.5 ft.

He then looked at what the effect would be of improving fuel consumption. He tried a specific fuel consumption of 2.0 and got a fuel weight of 36,100 lb, which would have left 16,000 lb for the airframe, slightly heavier than Lockheed's F-104A and one-third the weight of North American's proposed F-108. With that as a target, he began estimating the weights of airframe components. He assumed that the wings would be made of a steel waffle using 0.01-in.-thick stainless steel. This would have weighed about 2.16 psf, but to allow some room for error he assumed 3 psf.

Component	Area	Weight, lb
Wing	1400 ft²	5200
Tail	700 ft²	2100
Fuselage	48-in.-diam cylinder 80 ft long = 960 ft²	3000
Gear	——	1000
Nacelles	——	3000
Systems	——	1000
Payload and crew	——	1000
Engines	——	12,000
Total	——	28,300 empty

Johnson, C. L., Archangel project design notebook, Lockheed ADP, Burbank, CA, entry for 22 Apr. 1958. pg. 2.

This was only 300 lb over the estimate based on the optimistic fuel consumption, so he had reason to believe that he was on the right track.

Finally, he drew a sketch of an airplane with trapezoidal wings similar to those of the F-104, twin vertical stabilizers canted outward, engine nacelles merging into the fuselage, and movable horizontal stabilizers that could be rotated down, like those on the F8U Crusader (Fig. 34).

Three days later, Johnson called Gene Kiefer to report on the status of Project GUSTO. He reported Mel George's disappointing findings on the strengths of existing plastics, and his own conclusion that a composite aircraft would not be as strong as an all-metal one, and could not meet the range and altitude goals. He also described the aerodynamic problems with the GUSTO Model II design and how they were working on a variant with no tail, essentially a flying wing, but that it suffered from stability problems.

Kiefer later wrote,

> Kelly proposed to put Model II GUSTO on the back burner since it appears marginal at this time and take a look at another approach. Such

High Speed

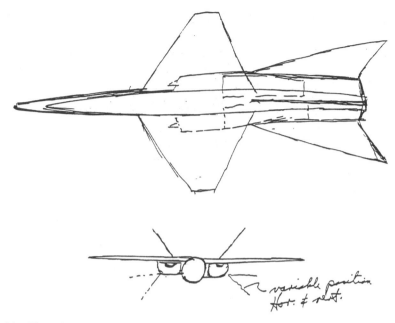

Fig. 34 First sketch by Kelly Johnson of a Mach 3 aircraft. (Courtesy of Lockheed Martin.)

approach would involve a configuration capable of $3\frac{1}{2}$ the speed of the AQUATONE vehicle and an increase of approximately 25% in altitude. Such RAINBOW protection as is available in the present state of the art would be incorporated in this design [57].

Johnson proposed to investigate the high-speed design for about 10 days, while continuing work on GUSTO II. Kiefer authorized this work and told Johnson to report his findings to Bissell not later than 16 May. Gibbs, who had also talked recently with Johnson, agreed with the plan.

Kiefer then led Johnson into a discussion of whether he should resume work on GUSTO Model I, which had not been designed with a minimum RCS as the overarching objective and thus might be more practical than the Model II: "Kelly was pessimistic, from information available to him, as to whether or not [Frank Rodgers] would be able to handle his end of the Model I GUSTO but agreed that this might be the only feasible approach" [57].

Kiefer discussed Kelly's ideas with Rodgers and suggested that he begin reexamining how the RCS of GUSTO Model I could be improved. He also described the proposed high-speed design and got a favorable response. When Kiefer wrote up all of this in a memo for Bissell, he recommended that Rodgers and perhaps Taylor attend the next meeting between Johnson and Bissell [57].

Jack Gibbs was now nearing the end of his time working at the Agency. In a memo to Bissell, he laid out the plan of operations for fiscal year 1959–60.

He summarized how he had separated the CHALICE (U-2) and GUSTO planning, after getting advice from Security on "how best to start and carry out the GUSTO program and not have it contaminated from birth by the erosion of security which has befallen CHALICE over the past three years" [58].

CHALICE he characterized as "not too clear"—probably because of the uncertainty of how many flights the President would allow—but the capability was clear enough that he could prepare a budget. GUSTO, on the other hand was "very foggy both as to technical feasibility and future mode of operation." Until they knew more about the technical possibilities, they could only do a little rough planning of costs [58].

In a mid-July memo to Dulles discussing the finances of his projects, Bissell was slightly more optimistic, saying, "The possibilities of this program looks [sic] most likely at this time from the feasibility studies conducted, but I feel it would be inadvisable to proceed on this program until all studies have been completed and we have had benefit of recommendations from the Advisory Panel ..." [59]. In a staff meeting three days later, he asked for one member to study a new report from OSI on Russian surface-to-air missile capabilities, in preparation for the 31 July advisory panel meeting. He felt that "this prediction on the Russian 1961 kill-probability might be overstated, but that such an estimate was crucial to proceeding with GUSTO" [60].

BLIP-SCAN STUDY

In the meantime, the SEI team had been working on an analysis of Johnson's proposal for a high-speed high-altitude aircraft. They began by analyzing where an aircraft would pass through a radar beam (see Fig. 35). Search radars of that era focused their energy in a beam between 2 and 14 deg above the

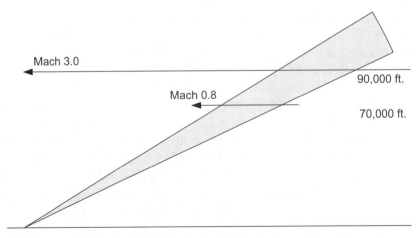

Fig. 35 Aircraft paths through radar beam (not to scale).

horizon. Below 2 deg the interference from ground reflections would cause too many false images. An aircraft flying directly at the radar would initially be below the beam. At 70,000 ft, the cruise altitude of the U-2, it would enter the bottom of the beam 170 miles from the radar set and exit the top of the beam at 70 miles. At 90,000 ft, Johnson's proposed U-3 would enter the beam 190 miles from the radar and exit it at 90 miles. "The problem at hand is to so construct an aircraft that its cross section during passage through the beam is so small as to be in the noise level of the radar in question ..."

To design an aircraft that could escape notice, they would have to know how many square meters of radar cross section a radar could see between 70 and 170 miles and between 90 and 190 miles for the two altitudes. Moreover, the analysis would have to be done for the three frequency bands used by different radars, 70, 600, and 3000 megacycles per second.

The first piece of good news was that the increase in range resulting from the higher altitude gave a 5 dB reduction in the strength of the reflection. However, the details of the analysis of antenna patterns and the corresponding minimum RCS numbers remain classified [61].

The scientists had to make many assumptions about the environment into which the aircraft would fly. The introduction of the paper reporting the results began,

> The calculation of the required performance for GUSTO is a dangerous occupation. It would be difficult even if we knew all the characteristics of all the [radar] sets we are facing for so many of the parameters involved in such a calculation are wildly fluctuating statistical factors. When you add to this the tremendous gaps of knowledge on what we face it almost becomes a farce to attempt an estimate [62].

A large part of the problem was figuring out how effective the Soviet radar operators would be. From their years at Lincoln Laboratory working on U.S. continental defense radars, they knew that in a test setting the radar is carefully maintained and the operator is a highly skilled radar engineer who is well rested and knows exactly when and where to look for the first blip from the approaching aircraft. They also knew that in military units operating and maintaining mass-produced radar systems that things would not be so efficient, or as they put it, "the human looking at a PPI is just not an optimal detector by several dB" [62]. Their problem was how to quantify the difference, so that they could give realistic objectives for the RCS of the vehicle.

The report summarized the analysis for several radars and concluded, "we could probably come up with a GUSTO version which could be quite effective against the present equipment" [62]. They believed that their assumptions were conservative enough that they would be valid as long as the Soviets were using mobile—rather than fixed-station—radars. Nevertheless, "I do not feel that we can afford to compromise very much on the performance represented by the Iron Maiden and would therefore recommend serious consideration of a flying wing version with a look at a stabilized mount for the payload." They concluded by proposing a list of measurements to be taken for the designs [62].

The analysis also showed the advantage of high speed. At a speed similar to the U-2's, the aircraft would take over 10 minutes to fly through the 100-mile-deep detection zone. However, a Mach 3 airplane would take about three minutes, increasing the chance that a radar operator would overlook the blips. And if the magnitude of the radar return were reduced, then the blips would be smaller and easier to miss.

The concept of combining speed, altitude, and stealth to avoid detection was elucidated in a report that came to be known as the "Blip-Scan Study." It set specific targets of a speed of Mach 3, an altitude of 90,000 ft, and an RCS of not more than 10 m^2 and preferably less than five. The concept also became known as the "Rodgers Effect."

Chapter 6

COMPETITION FROM CONVAIR

In the spring of 1958, Bissell brought a competitor for Lockheed into the project, the Convair division of General Dynamics. He flew to Fort Worth, Texas, and spoke with their head of advanced development, Robert H. Widmer.

Like Kelly Johnson, Bob Widmer was the son of immigrants; his parents had been born in Switzerland and had come to the United States during World War I. They returned for some years, and Bob attended elementary school in Switzerland. Back in the United States, he attended Rensselaer Polytechnic Institute in Troy, New York, and worked summers at Teterboro Airport, in New Jersey. Graduating during the depression, he could not find a job and decided to get a Master's degree from CalTech.

Bissell told Widmer that he needed a reconnaissance aircraft with a 4000-mile range, a 90,000-ft altitude, and a 2000-lb payload, and which could not be detected. As with Lockheed, Bissell kept the requirements simple; at first, there was no written specification for the aircraft that Convair was to develop (Interview with Robert H. Widmer, Fort Worth, TX, 13 March 2003).

Why did Bissell arrange for a competitor, when the U-2 had gone so well with Lockheed as the sole source? In his memoirs, he wrote, "Although Kelly Johnson was a close friend and Lockheed's Skunk Works had an exceptional track record, I felt I had to seek competing proposals" [63]. Bissell seems to have been motivated by a sense of due diligence and perhaps a bureaucratic need to protect himself from possible future criticism in awarding an enormously expensive contract.

Almost 40 years later Frank Rodgers remembered a different reason and described it in his blunt style:

> Bissell had a problem. His favorite airplane designer showed no interest at all in minimizing the radar cross section of any successor to the U-2. Instead, he seemed only interested in gathering further laurels for his Skunk Works by making an aerodynamic quantum leap. His experience with the dirty U-2 had convinced him that radar considerations would only limit his ability to realize a quantum leap. There would be no prestige attached to building another subsonic flying wing ten years after the originals had been cut up for scrap by the Air Force.

But the President was insisting that any successor be virtually invisible and this message apparently was not getting through to Kelly. Bissell decided his only hope was to find another contractor who could give Kelly some competition. (Rodgers, Franklin A., unpublished memoir, 3 Feb. 1995.)

In 2005, Leo Geary would neither agree nor disagree with Rodgers's view, but did allude to a "conflict" between Bissell and Johnson (Interview with Leo P. Geary, Denver, CO, 10 July 2005). Norm Taylor's recollection of events was the same as Rodgers's (Taylor, Norman H., e-mail to author, 25 Sept. 2003).

Whatever Bissell's reasons, he had come to the right place. While developing the B-58 bomber, Convair had created a lab for the study of radar cross sections and electronic countermeasures and had been attacking the problems in a systematic method. Convair had built a "Rat Scat" radar evaluation range in their "backyard" in Fort Worth and later replicated it at Holloman Air Force Base in New Mexico. They had worked with Lincoln Lab in the past and had reverse-engineered Soviet radars from their emissions and built working replicas for evaluation. The evaluations included man-in-the-loop simulations to understand the effectiveness of the complete radar system. Widmer has stated that in the course of their RCS investigations they had—independently of SEI—discovered the principle of a saucer's providing the minimum radar return.

Moreover, Convair, like Lockheed, was not afraid to try radical ideas. During the competition with Boeing to develop a supersonic bomber, the Boeing team had thought that Convair's entry would only be capable of Mach 1.2, like its own design. Unbeknownst to them, Widmer had decided to go for Mach 2. The Boeing team drew a sketch—completely wrong—of what they thought Convair's design would look like. During a trip to present the designs to the Air Force, they broke into Widmer's hotel room and left the picture, along with a poem, "Roses are red/Violets are blue/If your airplane looks like this/You're through!" Boeing was wrong, and Convair won with the B-58 (Interview with Robert H. Widmer, Fort Worth, TX, 13 March 2003).

Now, using the B-58 as a starting point, Widmer's group was developing a concept called the Super Hustler.

Super Hustler

Super Hustler [64] had grown out of Convair's work on the GEBO II (generalized bomber) studies for the Air Force. It used the B-58 to carry aloft a two-part parasite aircraft that would overfly hostile territory for reconnaissance or bombing missions. The front section of the parasite was a manned aircraft powered by ramjets for cruise and a turbojet for landing (see Fig. 36). Initially, it was propelled by a booster stage that contained two RJ-59 ramjets and, for bombing missions, a nuclear weapon. Made of stainless-steel

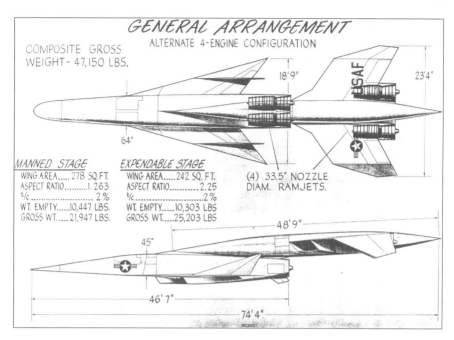

Fig. 36 General arrangement of Super Hustler manned and booster stages. (Courtesy of Roger Cripliver.)

honeycomb, Super Hustler was designed for flight at Mach 4, which would have created temperature that would have weakened any titanium structure (Fig. 37). With an additional rocket booster, it could be launched from a truck, avoiding the need for a B-58 or even an airfield. The purpose of the two-stage arrangement was to extend the range for the return flight (Interview with Robert H. Widmer, Fort Worth, TX, 13 March 2003).

Fig. 37 Super Hustler model. (Courtesy of Roger Cripliver.)

The manned stage of Super Hustler was designed for good crew coordination and protection. The pilot and "navbardier" (navigator/bombardier) sat side by side in a compartment that did not require the use of pressure suits, a factor that would reduce fatigue. To fit in the small vehicle, their seats reclined at 45 deg. During Mach 4 flight, insulated and cooled panels closed over the canopy and avoided the need to develop a glass resistant to the extreme temperatures. The crew looked outside using television cameras (see Fig. 38 for cockpit mock-up).

Although the vehicle was highly streamlined in cruise configuration, landing was a different matter. During the approach, the forward half of the vehicle drooped down 20 deg, and the protective panels folded up to allow limited vision for landing. A General Electric J85 engine located between the ramjets provided a flattening of the approach path, but not enough thrust to go around. The fuselage folded down behind the crew compartment so that the inlet would clear the ground, the tip of the nose folded up slightly to clear the ground, and the vehicle landed on two skids in the rear and one nose wheel (see Fig. 39).

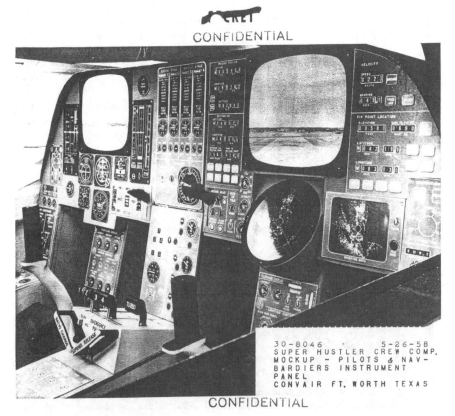

Fig. 38 Super Hustler cockpit mock-up. (Courtesy of Roger Cripliver.)

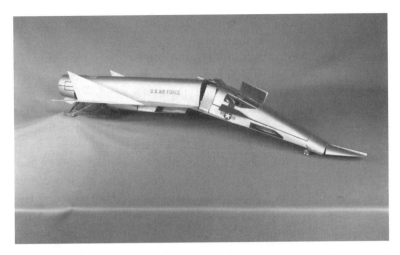

Fig. 39 Super Hustler manned component in landing configuration. (Courtesy of Roger Cripliver.)

A single complex inlet would feed both the ramjet and the turbojet. A splitter diverted the air left and right into ducts that led to the ramjets (see Fig. 40). The shape of the inlet created a series of internal shock waves that reduced the speed of the airflow while increasing its pressure. For subsonic operation

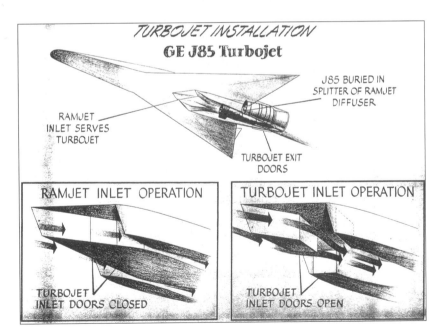

Fig. 40 Super Hustler inlet and duct arrangement had a splitter feeding rectangular ducts to the ramjets. Internal doors could redirect air to the J85 turbojet. (Courtesy of Roger Cripliver.)

using the turbojet, a door in each side of the splitter would open to admit air to a center duct. That duct led to the J85.

A major contributor to the Super Hustler propulsion system was Antonio Ferri, of the General Applied Science Laboratories, Inc. (ASL). Ferri was a world-renowned expert in high-speed flight. He was a professor at the Polytechnic Institute of Brooklyn and with the famous aerodynamicist Theodore von Kármán had founded ASL. It was at ASL that fundamental work on ramjet technology was done for Marquardt.

Super Hustler, however, was not designed for stealth. Its leading and trailing edges were all straight. The mounting of the manned component's wings against the sides of the inlet presented a long corner to reflect radar. The inlet itself would have created reflections. The twin vertical tails were mounted straight up, providing an easy target. And the booster component also had numerous straight lines, flat plates, and corners. Moreover, detaching the booster stage and letting it crash in denied territory would hardly have provided the unnoticed operation demanded by the President for peacetime.

FISH: FIRST INVISIBLE SUPER HUSTLER

Using the design principles from the SEI team, as well as their own understanding of the principles of stealth, a seven-man team under project engineer Donald R. Kirk began work on the First Invisible Super Hustler, or FISH.

Kirk's design team included one leader for each major area of the vehicle: G. C. Grogan, aerodynamics; A. E. Solis, electronics; C. L. Secord, control systems; R. C. Matteson, propulsion; Schreiber, secondary power and heat transfer; Collinsworth, airframe; and N. M. Alexander, structures.

Each leader had three to five others reporting to him. The engineers nominally worked for managers of departments who reported to Widmer, but to keep the number of people as small as possible the department managers were not involved in the FISH work. Kirk reported to Widmer, as did Vincent "Vinco" Dolson, who ran the model shop and manufacturing research groups. Security, contracts, and finance people also reported to Widmer. Ultimately, about 75 engineers worked on the Convair designs, and a total of 2000 were involved in all aspects of the project (see Fig. 41 for organization).

Rodgers enjoyed working with the Convair team:

> I found Widmer's organization much easier to work with than I had ever found Lockheed. They were eager and fast learners, as receptive to ideas from the outside as from the inside and did not appear to promote an agenda which differed from that of the customer. (Rodgers, Franklin A., unpublished memoir, 3 Feb. 1995.)

The main threat from radar was at low frequencies, about 90 MHz. The Convair radar studies had started with a disk, which would have given a minimum radar return from all angles. Like Kelly Johnson, Widmer and Kirk

COMPETITION FROM CONVAIR

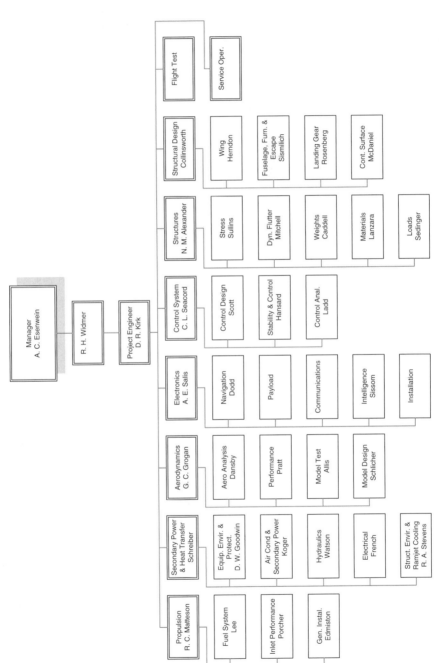

Fig. 41 FISH engineering organization. (Courtesy of the Jay Miller Collection and the Aerospace Engineering Center.)

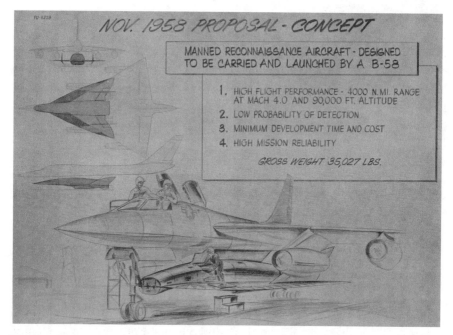

Fig. 42 FISH design of November 1958 with B-58 carrier. (Courtesy of the Jay Miller Collection and the Aerospace Engineering Center.)

were also unable to make a disk stable in flight, and so a somewhat more conventional design emerged.

The design was adapted from the manned stage of Super Hustler (Fig. 42). FISH would be launched from the B-58 and would fly its mission powered only by its own ramjets, which would burn high-energy fuel (HEF). To achieve the range with the added weight, the wing area was increased. To reduce the RCS, all of the leading and trailing edges were changed from straight lines to arcs of circles, and the inlet was redesigned. The steel honeycomb wings incorporated the wedge-shaped dielectric inserts invented by Ed Lovick. However, because of the higher heat that FISH would encounter at Mach 4, they were made of a ceramic, Pyroceram, impregnated with graphite. Because the coefficients of expansion of steel and Pyroceram were different, the wedges were designed to slip inside their mountings as the wings heated and cooled (Interview with Robert H. Widmer, Fort Worth, TX, 13 March 2003).

Unlike the Super Hustler manned component, FISH did not fold in the middle for landing, even though it retained skids for main landing gear and a small nose wheel. Removing that hinge saved weight and complexity and was probably made possible by the change to a shallower inlet. However, the nose of FISH still had to fold up to provide clearance for the B-58's nose gear. The size of the ramjets was increased from Super Hustler's 33.5 or 38.5 in. diam to

41.5. The J85 turbojet was replaced with a JT-12. The weight went up by more than 50%, from 21,947 to 35,027 lb. The weight available for the reconnaissance package was 560 lb, less than the 2000 originally requested by Bissell.

The inlet was completely different from Super Hustler's externally bifurcated design. A single wide and shallow inlet with rounded corners extended almost the width of the fuselage. A ramp from the belly slanted downward until it almost filled the inlet, compressing and slowing the incoming air, and reducing the radar return. Internally, the inlet branched left and right to feed two large ducts leading to the ramjets. A smaller duct branched off each large duct and joined to feed the turbojet. Because the point at which the turbojet's exhaust emerged was below the ramjets, it would have caused a pitch-up moment if the engine had been mounted horizontally. As a result, the engine face was placed higher than the exit, directing the exhaust downward and placing the thrust vector more nearly through the center of mass (see Fig. 43).

The two-man crew was reduced to one pilot who sat on the left side of the aircraft with a fuel tank to his right. As in Super Hustler, he did not look out of a canopy during flight, but relied on two TV cameras in the nose. The cockpit itself was a pressurized capsule that allowed the pilot to wear a flight suit and a standard mask, rather than the full pressure suit being proposed by Lockheed. Not wearing a pressure suit would greatly reduce the pilot's fatigue.

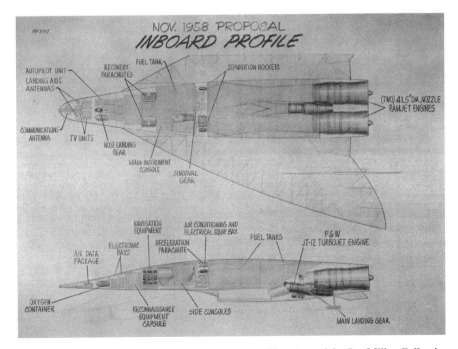

Fig. 43 Inboard profile of FISH showing systems. (Courtesy of the Jay Miller Collection and the Aerospace Engineering Center.)

The entire front of the aircraft forward of the inlet and the main fuel tanks was an escape capsule. Instead of firing an ejection seat in an emergency, the pilot would trigger rockets that would separate the capsule from the rest of the aircraft, and two parachutes stowed forward of the instrument panel would lower the entire capsule to the ground, a feature eventually implemented in General Dynamics's F-111. The all-important reconnaissance equipment occupied a bay forward of the parachutes.

A full-scale RCS model of FISH was tested on outdoor radar measurement ranges at Convair and at Indian Springs Air Force Base, using copies of Soviet radars. The radar return of the ramjet exhaust was also analyzed.

Bissell's office gave Convair's work the name Project IDIOM and issued a cost-plus-fixed-fee contract on 22 June 1958, which Convair accepted eight days later [65]. Because the work started before there was a formal budget, it was initially funded from Project CHAMPION, a Navy reconnaissance program. Then $79,000 was obligated from fiscal year 1959 funds [66]. Although the contract was managed under Project GUSTO, its documents were mostly numbered in the CHAM- series. Around the end of 1958, the work was formally moved into GUSTO.

By the end of August, Convair had ramped up their effort to the point where they needed to work overtime, which Bissell approved [67]. The competition was off and running. By November, Convair had finished its first version of FISH and was ready to present it to the Land Panel.

Chapter 7

ARCHANGEL I

"We called the U-2 the Angel. Obviously the follow-on airplane for it was an Archangel, right?"—Ed Baldwin

On 19 June 1958, one week after the Air Council had met to evaluate the final studies of liquid-hydrogen-powered aircraft, Johnson started a detailed study of a Mach 3 airplane, but one much larger than in his April sketches. His objective was to reach the performance goals with the least development risk; he wanted to deliver the aircraft in 18 to 24 months from the go-ahead. To that end, he would use a relatively conventional design.

He set a target of an empty weight of 41,000 lb, a gross weight of 100,000 lb, and a weight at the start of cruise of 88,000 lb. He computed a wing area of 1950 ft^2 and a span of 60 ft. He considered using a structure of either titanium or stainless steel.

When Gene Kiefer visited him on the 26 June, Johnson was worried about the engines. He wrote that "Kelly was not at all optimistic re either the design of an engine only for the high altitude high mach performance or the clobbering up of an existing engine" [68].

Kiefer apparently raised some possible features he had learned about the day before at Ramo-Wooldridge Corporation, such as improving heat resistance by putting ceramic coatings on engine components and using hollow turbine blades with coolant circulating inside:

> ... the J-58 operates at a turbine inlet temperature of 1850°. If this could be raised the engine's specific fuel consumption could be considerably improved and perhaps the after-burner eliminated. We batted around the engineering problems and the political aspects of the problem so as not to dampen Kelly's enthusiasm and yet overcome the undesirable features of his supersonic machine. He expressed an interest in trying to combine supersonic speed with the electrical properties of [Frank Rodgers's] configuration which was something he had not attempted to do previously [68].

Back at his desk, Johnson did a budget estimate for the work. Dick Boehme had 40 men, including himself and Henry Combs. Using a figure of $10 per hour, one week would cost $16,000. He added a profit of $500 per week and concluded, "For $225,000 can go full steam for 3 months" (Johnson, C. L., Archangel project design notebook, Lockheed ADP, Burbank, CA, entry for 26 April 1958, pg. 1). With $200,000 for wind-tunnel testing and $100,000 to build a mock-up, the total bill came in at $525,000; $808,000 had been allocated for GUSTO, of which $355,000 was unspent and available. This left him $170,000 short, and so he decided to get $175,000 to complete the funding for work from 15 July to 15 October.

Johnson next did an evaluation of the J58 with and without afterburners, as well as a weight breakdown of the airplane (see Fig. 44). He concluded that when half the fuel had been burned, the afterburning airplane could hold an altitude of 86,000 ft, whereas the nonafterburning one could only reach 68,000 ft. He focused on the afterburning model and concluded that it would have an operational radius of 1600 n miles, which was 400 miles less than the Agency required. He looked at reducing the structural weight to 41,000 lb, which would allow 43,000 lb of fuel at cruise, and consulted Dick Fuller, who estimated that the climb to altitude would take 268 miles. Added to the

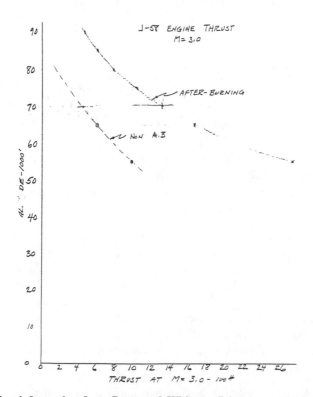

Fig. 44 Using information from Pratt and Whitney, Johnson graphed the expected thrust at Mach 3 of the J58 for various altitudes. The afterburner more than doubled the total thrust. (Courtesy of Lockheed Martin.)

other ranges he had recomputed, this gave a radius of 2000 n miles and was "Our only chance."

For the breakdown of the 41,000 lb, Johnson worked with Merv Heal and Dick Boehme. They came up with 24,685 lb for the airplane without engines, 800 lb for the crew and payload, and a maximum of 15,515 lb for the engines. This meant that they could not afford to hang the engines from pylons and would have to put the engines in the fuselage. They even looked at putting 3250 lb of fuel in the vertical tail to squeeze out a bit more range.

Johnson went to his drafting board and drew a 1/60th-scale drawing of the aircraft (Fig. 45). He gave the drawing to Ed Baldwin and other members of the team. Baldwin "looked it over with Dick Boehme and some of the other guys and we really couldn't hang onto that tail. ..." But the drawing was "a good starting point and that's all that we wanted" (Baldwin, E. P., oral history, unpublished audio recording).

The "other guys" included the members of a small conceptual design team. As usual, Dick Boehme was the project engineer. The others were Henry Combs, structures, stress analysis, and configuration; Ed Baldwin, configuration layout; Dan Zuck, configuration layout, cockpit integration; Ben Rich, propulsion and thermodynamics; Elmer Gath, powerplant; Dick Fuller, aerodynamics; Ed Martin, reconnaissance systems; L. D. MacDonald, electromagnetics; and Ed Lovick, electromagnetics.

On 3 July, just over a week later, Baldwin had the GA drawing finished for Archangel I (Fig. 46). It had a 1650.15-ft^2 wing, a 388.35-ft^2 vertical tail, a 390-ft^2 horizontal tail, and J58 engines in nacelles adjacent to the fuselage. The majority of the structure used a titanium alloy, B120CVA. (See Fig. 47 for the model of Archangel I.)

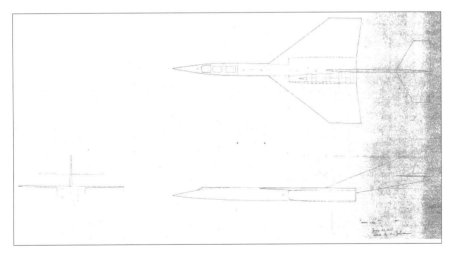

Fig. 45 Archangel first concept drawing, by Kelly Johnson. (Courtesy of the family of Edward P. Baldwin.)

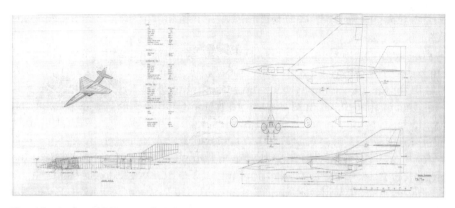

Fig. 46 Archangel I general arrangement drawing, by Ed Baldwin. (Courtesy of the family of Edward P. Baldwin.)

Archangel I was capable of a 4022-mile mission at Mach 3.0 (see Fig. 48). It would take 306 miles from takeoff to reach a cruise altitude of 83,000 ft. As fuel burned off, it would become lighter and would climb. By the midpoint of the mission, it would be at 88,000 ft and would be up to 93,000 ft before it had to begin its descent. Upon landing it would have a fuel reserve of 2000 lb in its tanks.

Wind-tunnel testing found a problem that could have prevented reliable—or at least efficient—engine operation. The corner where the wing met the fuselage generated turbulent airflow into the inlet. As the aircraft maneuvered—rolling, banking, and yawing—the flow would have varied, possibly causing inlet unstarts.

In an attempt to improve the altitude, ramjets were added to the wing tips. However, because there was no corresponding increase in wing area, the

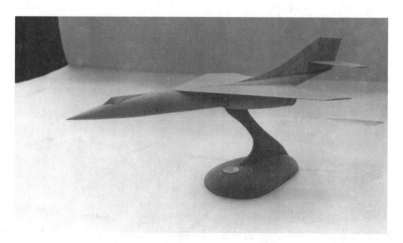

Fig. 47 Archangel I desk model. (Courtesy of Lockheed Martin.)

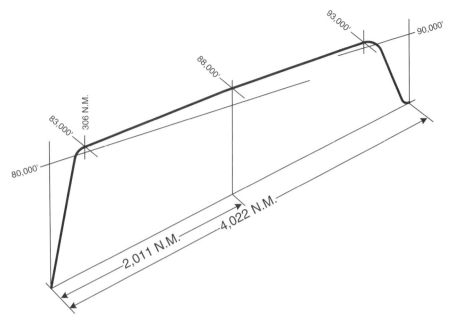

Fig. 48 Archangel I mission profile. (Drawn by the author from Whittenbury.)

ramjets increased the wing loading. That, together with the increased fuel consumption, decreased the range.

At about this time, Johnson obtained basic data on the proposed F-108 Mach 3 fighter, which was similar to the Lockheed design in size and weight. He included sketches of the F-108 in his design notebook. A number of CIA historians have seen these sketches and, not recognizing the F-108, described the drawings as a version of the A-1.

On 23 July, Johnson presented both Archangel I and GUSTO 2A at the Cambridge office, with Bissell, Kiefer, Donovan, Purcell, and H. M. "Dick" Horner, of United Aircraft, in attendance. Johnson wrote that both designs were "well received." A Navy Commander Dewey Struble told him of the Navy's idea for an inflatable airplane. Bissell wanted Johnson's comments on the concept.

Eight days later, on 31 July, the advisory panel met to discuss the Lockheed concepts, which they called GUSTO-A and GUSTO-B, and probably the Navy concept. Although no conclusions were reached, Bissell told his staff on 5 August that he felt that reasonable progress was being made and that the path forward would become more clear after the September advisory panel meeting. One member of the staff suggested contacting the Air Technical Intelligence Center (ATIC), at Wright Field, for a new estimate of Soviet interceptor capabilities against the presumed performance of the follow-on vehicle. Bissell agreed, but insisted that the capabilities of the follow-on not be made too specific, in order to maintain security [69].

THE IRON MAIDEN

Frank Rodgers had been doing further work on shaping the aircraft for a broadband solution:

> I had had a full scale model built of a section of the fuselage Kelly was proposing for his new design. But instead of the simple cylindrical fuselage Kelly was proposing, I "stretched" the sides of the cylinder so the cross section resembled the cross section of a rather fat flying saucer or the cross section of a racing boat, complete with chines. The "broad-band" treatment which I had discovered before leaving Lincoln was applied to the chines. Lockheed personnel rather derisively referred to my model as the "Iron Maiden." As soon as data was collected on the radar cross section of this model, it was fed back to Lockheed. Nothing was held back. (Rodgers, Franklin A., unpublished memoir, 3 Feb. 1995.)

The Iron Maiden cross section turned out to be one of the enduring results of Project GUSTO. Not only was it incorporated in the final design of the successor to the U-2, but similar shaping was used in other military aircraft over the next half-century. Besides reducing RCS, chines can also improve stability in high-speed flight, a benefit that Rodgers probably did not foresee, but which was realized by Lockheed.

FUNDING FOR GUSTO

On 12 August, Bissell went to his boss with a request for a funding for his various reconnaissance projects, including GUSTO:

> During the past year, the activities for which I have been responsible as the Director of Project AQUATONE have multiplied. Certain new tasks were handled as subprojects of AQUATONE without formal approval by you as separate projects, and with no separate funding or accounting. Others were handled in an ad hoc manner as new projects but with approval by you of only the sums initially provided therefore. It appears desirable in the current fiscal year to handle these several tasks as separate projects [70].

Besides GUSTO, the work included U-2 operations (now renamed from AQUATONE to CHALICE) and the spy satellite development under Project CORONA. Although the Agency had submitted its regular budget for fiscal year 1959 to Congress, Bissell's projects had expanded rapidly in scope, and the costs had gone up accordingly. In the case of GUSTO, the money went not just for models and mock-ups by Lockheed and Convair, but also for radar measurement ranges in the western United States [71].

Bissell proposed that for the next six weeks the projects be funded from Project CHALICE funds; the expected costs would be within the amount that Congress had approved. By 1 October, the scope of the other projects would be more clear, and the Bureau of the Budget would be requested to release funds from the Agency Reserve. Allen Dulles approved Bissell's request.

Chapter 8

NEW IDEAS

CHAMPION: NAVY INFLATABLE DESIGNS

On 14 August 1958, three weeks after his previous trip, Johnson was back at the program office to get the details of the Navy proposal. Under a Navy project named CHAMPION, Goodyear was proposing a reconnaissance vehicle having inflatable wings that could be rolled up while the vehicle was transported on an aircraft carrier and then inflated for launch (Interview with Sherre Lovick, Northridge, CA, 4 Feb. 2006). It was intended to be ramjet powered and to cruise at 125,000 to 150,000 ft. A balloon would lift it to altitude. Johnson made a quick calculation and decided that the balloon would have to be over a mile in diameter. He is said to have remarked, "Gentlemen, that's a lot of hot air."

Goodyear's propeller-driven Inflatoplane had already been flying for two years, but at a maximum speed of 72 miles per hour and at a maximum altitude of 10,000 ft. The Land Panel probably recognized that Goodyear was in a situation similar to Randolph Rae and Garrett, a builder of a small, simple aircraft proposing to build a large, complex aircraft to operate in a flight regime in which the company had no experience. Having an experienced company like Lockheed perform a sanity check on the concept was essential.

Despite his skepticism, Johnson accepted the task and on 18 August outlined the study in his notebook. His team would evaluate a variety of launching techniques, structures, and performance. To put a stake in the ground, he assumed a basic vehicle weight of 20,000 lb, a range of 4,000 n miles, an altitude of 150,000 ft, a payload of 800 lb, and a wing with a L/D of 5.0. From that they would derive the wing weight, the number and size of ramjets, and the size of the balloon. Besides hoisting the vehicle with a balloon, they would also look at using another airplane to tow the vehicle to 80,000 ft, at which point a rocket booster would accelerate the vehicle to ramjet ignition speed and at least 120,000 ft. A third launch concept was to use a low-thrust liquid-fueled rocket to launch from the ground to about 100,000 ft altitude for start of ramjet-powered flight.

The next day Baldwin began designing the "tug" aircraft to tow the reconnaissance vehicle to altitude. It had to tow a 25,000-lb aircraft with a L/D of

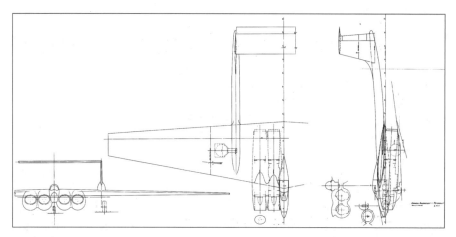

Fig. 49 Peterbilt tow plane general arrangement drawing, by Ed Baldwin. (Courtesy of the family of Edward P. Baldwin.)

10 (he apparently felt the higher L/D was justified) to 70,000 ft at a Mach number of 0.8. Three days later he completed the three-view drawing.

Baldwin appears to have reused at least some of the structural work from GA #2 because it also had straight wings and a horizontal tail supported above twin booms. Because there was no need for stealth, its four J75 turbojets were placed under the wings and were clustered adjacent to the small fuselage. Because of its huge size—the wing span was 110 ft—the tug became known as "Peterbilt," after the truck (Fig. 49).

Johnson evaluated three designs for the reconnaissance vehicle (Fig. 50). Configuration A used inflatable structures for the wings and tail. B used a short ramjet slung underneath a continuous wing. C was constructed all of metal and had the wings mounted at midfuselage of a long engine whose diameter was 180 in.—15 ft. The cockpit would be placed in the centerbody of the ramjet inlet. Although that was unusual, it was not unprecedented; one concept for Convair's XP-92 used the same arrangement with a turbojet

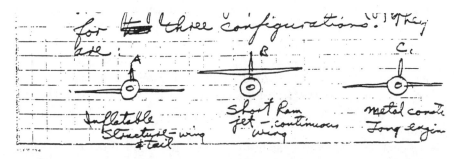

Fig. 50 Project CHAMPION design sketches, by Kelly Johnson. (Courtesy of Lockheed Martin.)

engine and resembled one of the Navy designs, although it never went past the mock-up stage. Similarly the turbojet-powered British Miles M.52 would have placed the cockpit in the inlet's shock cone. By the last week of August, Dan Zuck had produced three-views of the designs. (See Fig. 51.)

Two surfaces of balloon material were held together with millions of little threads, a construction technique that the engineers called a "fur burger." Ed Baldwin described it as an adiabatic airplane because the higher the internal pressure, the higher the load factor the aircraft could provide (Baldwin, E. P., oral history, unpublished audio recording).

Through the first week of September, Johnson wrote and dictated his report, "Evaluation Studies of Inflatable High-Altitude Aircraft," which would be designated SP-100 and published on the 11th. The final conclusions were as follows:

1) Inflatable wing and tail surfaces for the proposed design cannot be built for 1 psf for the speeds and altitudes proposed.

2) A metal surface can be built for 80% of the weight of the inflatable surface for the same speeds, altitudes, planform, and thickness. It would however still be 20% over the desired unit weight.

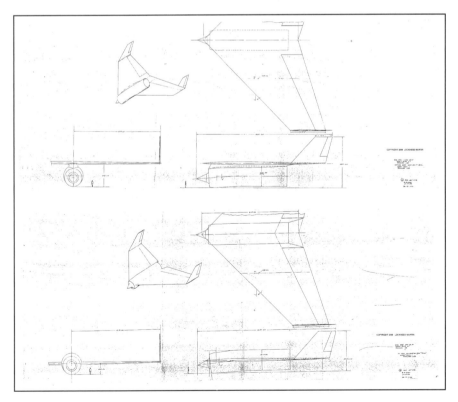

Fig. 51 "Ram Jet Kite" three view drawings, by Dan Zuck. (Courtesy of Lockheed Martin.)

3) Ramjet powerplants will operate at the speeds and altitudes proposed when run on Borane fuels.

4) Even using the most optimistic design criteria in terms of speeds, gust loads, structure, drag, and propulsion, no aircraft system having any reasonable degree of feasibility could be derived to fly the desired mission previously outlined.

5) The best launching means for the aircraft studied was a combination of towing by another aircraft and then boosting to speed and altitude by rockets.

6) The great technical risks involved, high cost, great vulnerability, and overall lack of feasibility for the size and weight proposed would indicate that other approaches to the problem should be considered more fruitful.

Goodyear's Inflatoplane project continued for another 15 years, but only as a small, simple airplane; it apparently was never again considered for the strategic reconnaissance role.

ARCHANGEL II

Johnson's next objective was to revise Archangel I to get the same 4000-mile range, while flying 10,000 ft higher, regaining the altitude lost in the transition from hydrogen to hydrocarbon fuel. This would require less wing loading (more wing area and/or less weight) and much more thrust. Apparently inspired by the Navy concepts he had been asked to evaluate, Johnson began thinking about a new direction for his own designs almost immediately after the meeting. On 18 August 1958, he made two notes "On Archangel: a. Put on ram-jet tip tanks ... b. In general terms—what does staging do? Consider lower design wt. by amt. of climb fuel." Using ramjets on the wing tips of the F-104 had been the subject of a proposal he had made to the Air Force in 1954. (Johnson, C. L., Archangel project design notebook, Lockheed ADP, Burbank, CA, entry for 18 Aug., 1958, pg. 4.)

Over the next week he made notes on Lockheed's experience with the X-7 ramjet test drone and got ramjet data from Ray Marquardt. Considering how long Archangel II would have to cruise on ramjet power, actual ramjet flight time was lacking. In about 100 flights of the X-7, Lockheed had accrued only about three hours total time, less than two minutes per flight. Adding the experience of Boeing and another manufacturer, the total time of all ramjet flights was only about 10 hours. Vibration seems to have been a problem, with the test vehicles running "rough as a cob" with up to 50 Gs at times. What problems would surface when a ramjet ran for not just a few minutes per flight, but for the two hours necessary to cover 4000 miles? Reliability would be essential for an airplane 2000 miles from friendly territory, where even a partial power failure could leave it at a lower altitude where it could be detected and shot down.

With basic data in hand, on the 27th Johnson began writing up "A Study on Getting Archangel II to 100,000' Cruise Altitude by Going to $M = 3.2$ plus

Ram-Jets on tip." He felt that ramjets would add 10,000 ft to the cruise altitude of Archangel I. He would increase the wing area from 1650 to 2500 ft^2, the tail area from 780 to 1200 ft^2, and the empty weight from 41,000 to 50,000 lb. With a gross weight of 100,000 lb—about the same as Archangel I—half of the weight would be fuel, and the aircraft could deliver a cruise range of 4000 n miles, plus 400 miles for climb and descent. The disadvantage was that it would be burning pentaborane, a toxic fuel. Using the fuel was justified because, "It has been shown that borane burns better at high altitude than even liquid hydrogen" [72].

Ben Rich, Dick Fuller, and Don Nelson provided drag and thermodynamic information and confirmed Johnson's estimates. He turned the design over to Ed Baldwin, who produced a general arrangement drawing on 3 September (Fig. 52). By the 11th they had completed a 15-page report, SP-101, "Design Study: Archangel Aircraft," presenting the two designs [73].

By all measures, Baldwin's design for Archangel II was the largest of all of the aircraft proposed to the Agency (CIA). The length was 129.17 ft, the wing span was 76.68 ft, the height was 27.92 ft, and the takeoff gross weight was 135,000 lb. It also had the highest cruising altitude, as much as 105,000 ft. It had modified biconvex delta wings with a L/D ratio of 6.2. They were mounted flush with the top of a cylindrical fuselage, a J58 turbojet under each wing, and a 75-in.-diam ramjet at each wing tip. The larger wings allowed moving the J58 nacelles away from the fuselage, which reduced the bending moment on the wings and thus a lighter wing (see Fig. 53). The J58 engines would be modified to allow the extra 0.2 Mach number, at a cost of 140 lb weight.

The combined turbojet plus ramjet powerplant allowed the aircraft to use the high efficiency of a ramjet at cruise while avoiding the takeoff and landing difficulties of a ramjet-only vehicle, that is, needing a boost to ramjet ignition speed and having to land deadstick because ramjets would not work at low speeds. A mission would see the ramjets lit at 36,000 ft and Mach 0.95. Because their specific fuel consumption was high below about Mach 1.6, a removable nose cowl would have to be added to get reasonable thrust at those speeds. The J58s would burn JP-150, a variant of the standard JP-4 fuel.

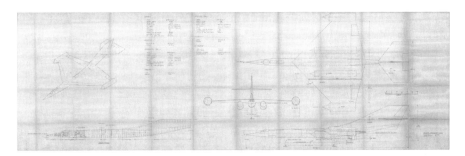

Fig. 52 Archangel II three-view drawing, by Ed Baldwin. (Courtesy of the family of Edward P. Baldwin.)

Fig. 53 This model of Archangel II places the turbojet nacelles farther outboard than in Baldwin's GA drawing. The GUSTO 2 wing is shown for comparison. (Courtesy of Lockheed Martin.)

The ramjets used required no great breakthroughs, and Marquardt estimated that they could be in service within two years. The decision to use ramjets that were straightforward adaptations of existing ones meant that they would weigh 2000 lb each. Lighter ones might have been possible, but there would have been a greater risk to the schedule.

SFC varied from 7.7 lb of fuel per pound of thrust per hour at Mach 0.95 at 36,000 ft, down to 3.2 lb of fuel per pound of thrust per hour at Mach 3.2 at 100,000 ft. As the aircraft became lighter, the ramjet fuel mixture would be leaned back so that the SFC would be down to 1.6 at 105,000 ft (see Fig. 54).

Unfortunately, Archangel II had no provisions for reducing RCS. There were corners where the wing met the engine nacelles, where the wing met the fuselage, and where the horizontal and vertical tails met. The circular inlets and nozzles for both the J58s and the ramjets would produce significant returns. Johnson was about to hear again that the President wanted an aircraft that was invisible.

WEEKEND WORK: A SMALL AIRPLANE

Johnson spent the week of 17–24 of September 1958 in Washington and Boston. Perhaps during his visit to the Agency on 18 and 19 September he was told that Archangel I and II were too large because on Saturday, 20

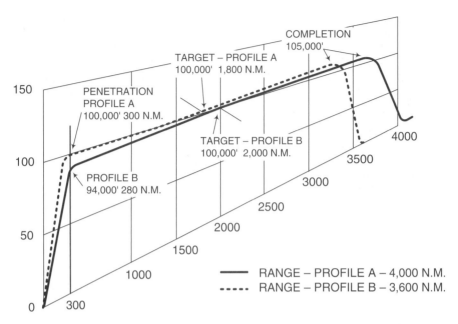

Fig. 54   Archangel II mission profile. (Drawn by author from Lockheed SP-101.)

September—two days before meeting with the Land Panel—he began making notes on a much smaller airplane that could cruise at higher altitudes. He began with an analysis of a lightweight ramjet proposed by Pratt & Whitney (see Fig. 55). He decided that the numbers he had been given for the ramjet must be wrong because they added up to the same performance as the earlier SRJ-54 ramjet, but at an altitude 40,000 ft higher than the SRJ-54 could do. He called Perry Pratt who was unable to explain the discrepancy, but admitted to Johnson that the numbers were "optimistic."

Johnson then dove into the design of the new airplane (Fig. 56). It would have a metal skin, be powered by ramjets burning SF-1 (liquid hydrogen), and would fly Mach 3.0 at 135,000 ft, possibly in an attempt to get it even higher above Soviet radars. However, this would be a much smaller airplane because at ramjet ignition the total weight would be a mere 10,000 lb. He calculated that he needed a wing area of 1670 ft^2 and that to keep it light would require a wing made not of all metal, but of duralumin and fabric. That would probably give 1.1 psf, but the top speed would be only Mach 2.5. To get back to Mach 3, he considered using only titanium. In the end he had a 100-ft-long airplane with a 50 wing span that looked very much like Archangel 1, but that "No useful range is available at this wt."

He then went to a heavier design, where the 10,000-lb weight did not include fuel. The airplane could carry 5000 lb of SF-1 fuel (liquid hydrogen), but it would require 700 lb of insulation for the tank, and "All the CL-400

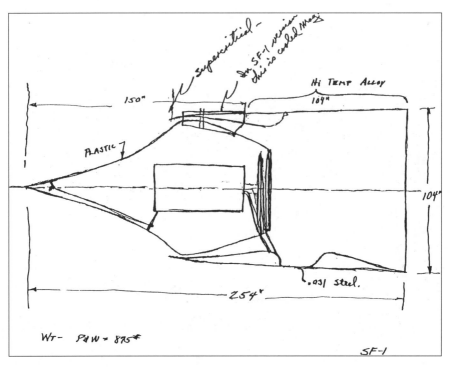

Fig. 55 Ramjet inlet sketch, by Kelly Johnson. (Courtesy of Lockheed Martin.)

problems come up—tank expansion, etc. Logistics." And the best range he felt he could get would be 2960 n miles.

So far the designs weren't practical, so Johnson decided to "Try a Borane Job" using high-energy fuel to propel the 10,000-lb airplane to 135,000 ft. This design would keep the same wing area, 1670 ft^2, and would "eliminate

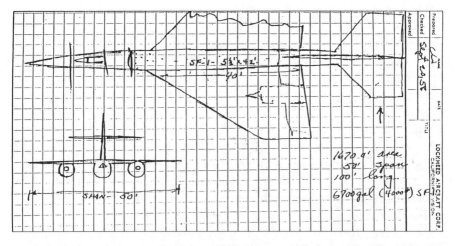

Fig. 56 Small ramjet-powered design, by Kelly Johnson. (Courtesy of Lockheed Martin.)

the fuselage except for the cockpit and equipment bay." The length would be 75 ft and the wing span 50 ft. The single vertical stabilizer would be replaced by two, one at the tip of each wing, resulting in a planform vaguely like a supersonic, delta-winged version of GUSTO 2A but with engines under the wings (see Fig. 57). He calculated that the weight would be 7960 lb, leaving room for 2000 lb of fuel. Unfortunately, the range was only 1090 n miles. Reducing the cruise altitude to 125,000 ft bought a little improvement in range—to 2520 n miles—still far short of the required 4000.

On Sunday, 21 September, Johnson tried an even more complex design, a 15,000-lb airplane with a 14,000-lb booster rocket (to get it to ramjet ignition

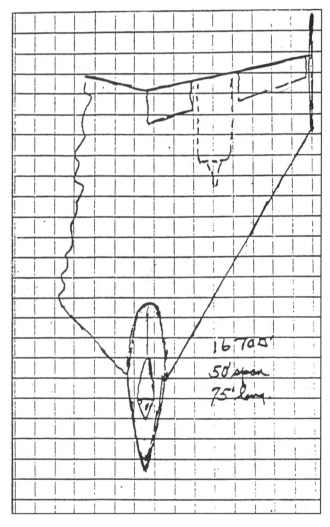

Fig. 57 Small aircraft design without empennage, by Kelly Johnson. (Courtesy of Lockheed Martin.)

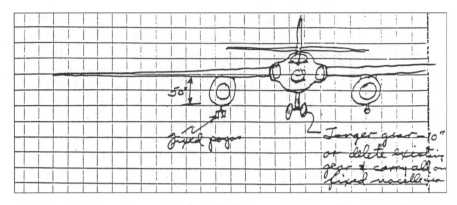

Fig. 58   U-2 adapted for towing, by Kelly Johnson. (Courtesy of Lockheed Martin.)

speed), all towed to altitude by a modified U-2. The existing U-2 had about 1000 lb of excess thrust at 50,000 ft and full weight. By hanging an afterburning J75 under each wing, the tow plane could fly at 60,000 ft with 5000-lb extra thrust for towing the reconnaissance aircraft and booster. He concluded that the "Proposed U-2 is perfectly feasible and cheapest aircraft available to provide tow at low speeds and with sufficient excess thrust to get ram-jet vehicle to 60,000′ or more at $M = .80$" (see Fig. 58). (Johnson, C. L., Archangel project design notebook, Lockheed ADP, Burbank, CA, entry for 21 Sept. 1958, pg. 14.)

LAND PANEL

On 22 and 23 September 1958, both Johnson and Widmer were in Boston to present their designs to the Land Panel. The members of the panel were from both government and industry. In addition to Land, Purcell, Stever, and Rodgers, the Agency was represented by Bissell and Kiefer. From the Air Force were General Swofford, Colonel Norm Appold, Colonel Seaberg, and Major Bob Hippert. Struble of the Navy returned, as did Horner of United Aircraft, and one Brady of Convair participated for an hour.

Widmer, accompanied by Kirk and the Convair radar experts, made his first presentation of FISH. Although each company was given some information on the other's work, Widmer recalled that he and Johnson never saw each other at any of the meetings, although he was sure there were times when they had both seen the panel on the same day.

Johnson had a large amount of data to present. He talked about his evaluation of the Navy designs, GUSTO 2A and Archangel II. He apparently delivered two reports, SP-100 comparing the Navy designs and SP-101 on Archangel I and II. Johnson recorded in his log that GUSTO IIA was very well received by the Land Panel, but that Archangel II was rejected because of its use of penta-borane for the ramjet and its high overall cost [73].

The panel apparently also raised questions about ramjet reliability and objected to the lack of attention to RCS.

The panel also gave Johnson information on Convair's FISH and asked that he proceed with an evaluation of its practicality.

There was also discussion of the Boeing proposal for a hydrogen-powered aircraft weighing 167,000 lb and being 200 ft in length, which had been done for the Air Council. This was apparently the last discussion of a SUNTAN design by the panel [74].

At a staff meeting on 25 September, Bissell summarized the meeting for Burke, Geary, Reber, and others. He said that

> ... there had been a good exposure of services personnel and scientists to each other at these meetings. No final recommendations had been reached because of the necessity for more investigation in one particular program. The meetings succeeded in eliminating present consideration of the [CHAMPION] program, whose timing developed to be out of place with the hoped-for schedule for any follow-on program. Mr. Bissell stated that another meeting of the Advisory Panel would take place in late October or near the first of November. [75]

The program that needed further development was apparently Lockheed's because Johnson wrote that, "We left Cambridge rather discouraged with everything." On the way back to Burbank he began designing the new airplane in earnest. (Johnson, C. L., A-12 Log, Lockheed Martin, entry for 17–24 Sept. 1958, pg. 1.)

THE A-3

> On the way home, I thought it would be worth a try to break one existing ground rule—namely, that we should use engines in being. It was this factor which made the Archangel II so large, as we started out with some 15,000 to 18,000 pounds of installed power plant weight on the J58's alone. Because the JT-12A is a low pressure ratio engine, it seemed to me to be well-suited to high Mach Number operation. I made a few numbers trying to scale down Archangel II to the 17,000 to 20,000 pound gross weight, and it appears feasible [76].

Although Archangel II might have been rejected because of its cost and its use of a problematic fuel, its radar return must also have been an issue because five days later Johnson began his notes on the A-3 and wrote that the basic concept was to reduce RCS. He proposed powering the vehicle with two JT-12As with afterburners in the fuselage and two 30-in. ramjets on the wing tips, reducing the payload to 300 lb, and aiming for Mach 3.0 or 3.2 at 100,000 ft. The design had no horizontal stabilizer in an attempt to reduce

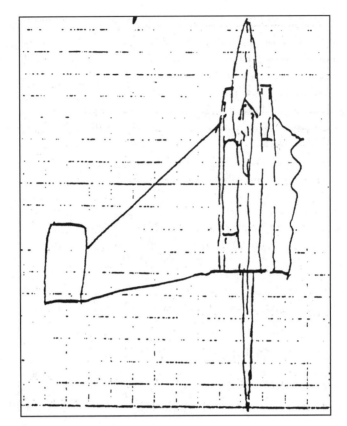

Fig. 59 First A-3 sketch, using ramjets and turbojets, by Kelly Johnson. (Courtesy of Lockheed Martin.)

the RCS, as well as to reduce weight (Fig. 59). His first sketch bore no resemblance to the work he had done while in Washington.

Another sketch used a single 75-in. ramjet in the fuselage, with the two JT-12A turbojets on either side (Fig. 60). The wing had a much more acute sweep, and there was a vertical stabilizer almost at the tip of the wing. Outboard of the vertical on each wing was an elevon.

The main problem with the second design—essentially a flying wing— was that it did not have enough room for fuel; a fuselage would be needed after all.

Johnson soon added some more goals to the low RCS objective:

1) The aircraft should be capable of using boron fuels.

2) Initial cruise altitude should be 90,000 ft and altitude at the target 95,000 ft.

3) Radius of action at Mach 3.0 to 3.2, including a 180-deg turn, should be 1500 n miles with hydrocarbon fuel and 2000 n miles with borane fuel.

4) The aircraft should have minimum weight and cost.

NEW IDEAS

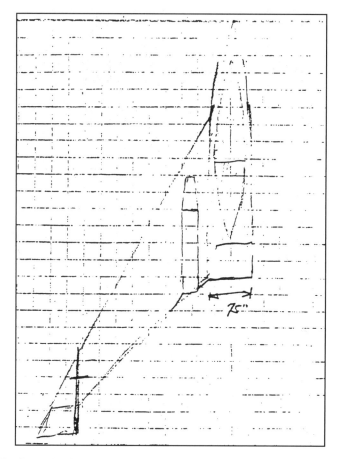

Fig. 60 Second A-3 sketch, by Kelly Johnson. (Courtesy of Lockheed Martin.)

On 3 October 1958, Johnson wrote up a page with the weight goals and gave a copy to each member of the team. If everyone could bring in his piece of the design at the target, the total empty weight would be 9000 lb.

CHERUB

Johnson assigned the first concept to Dan Zuck, Ed Baldwin, and Henry Combs. They called their designs Cherubs, in keeping with the angel theme and because they were small. All had ramjets at the ends of the wings and JT-12As in the fuselage. The ramjets had elliptical inlets, which would have helped to reduce the radar return. Baldwin's Cherub #1 had slight dihedral (Fig. 61), while Zuck's Cherub #2 had more acutely-swept wings with significant anhedral. Combs's design moved the turbojets on top of the wings (Fig. 62). The designs did not have a horizontal stabilizer, removing one source of a corner to reflect radar, although the large vertical stabilizer would have provided a large

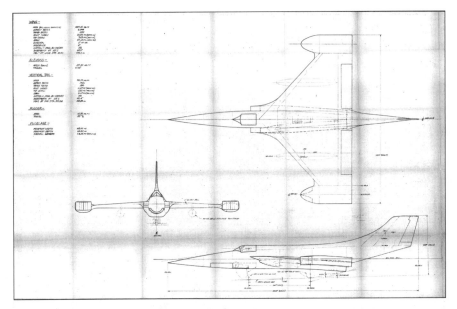

Fig. 61   Cherub 1 (A-3) three-view drawing, by Ed Baldwin. (Courtesy of the family of Edward P. Baldwin.)

reflection to a radar located to the side. The equipment bay was placed forward of the cockpit, and on the front view its outline looks like a smile.

Final Design

The final A-3 design, completed by Baldwin on 30 October, was similar to Comb's design (Fig. 63). The 40-in. ramjets remained at the wingtips, and the rectangular inlets were replaced with circular ones.

The main factor in keeping the A-3's RCS low was its small size. Placing the JT-12As above the wings minimized the corner between the wing and the nacelle. However, there were still corners where the wings met the fuselage and where they met the ramjets. Moreover, the inlets of all of the engines projected forward of the wings, providing a good radar target. The design was still small. The length was 62.3 ft, and the span was 33.8. The zero fuel weight was 12,000 lb, and fuel was 18,000, almost 2/3 of the takeoff gross weight. Wing area was a mere 500 ft^2. Nevertheless, it could fly 4000 n miles at Mach 3.2 and reach a cruise altitude of 95,000 ft.

The relatively low thrust of the JT-12As meant that extreme weight control had to be exercised, so that the turbojets could get the airplane to an altitude and speed where the ramjets could be lit. Wind-tunnel tests were very limited, and there were some hope that if transonic drag was less than initially expected, then less fuel would be needed to get to ramjet ignition and the available weight could be used elsewhere.

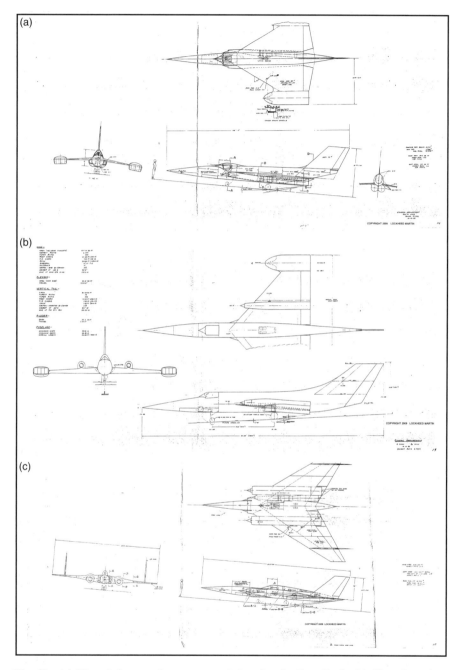

Fig. 62 (a) Cherub 2 general arrangement drawing, by Dan Zuck. (b) Cherub variant general arrangement drawing, by Henry Combs. (c) A-3 variant general arrangement drawing, by Dan Zuck. (Courtesy of Lockheed Martin.)

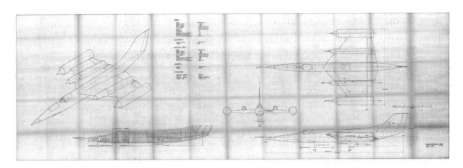

Fig. 63 Final A-3 three-view drawing, by Ed Baldwin. (Courtesy of the family of Edward P. Baldwin.)

WEIGHT REDUCTION

There were many subtleties to the design, most of which were driven by the small size. One of these was actually a bit of wishful thinking, namely, that the equipment could be reduced below that of the U-2, to a target in the range of 250 to 300 lb.

The fuel system was designed to squeeze as much fuel as possible into the small volume. The entire fuselage behind the cockpit was a fuel tank; this meant that the equipment bay had to be in front of the cockpit. The wings carried fuel to be burned in the climb. This eliminated the need to cool this fuel, as it was burned before the aircraft reached maximum temperature; none of the fuel was expected to heat beyond 250° F. If later analysis proved that to be incorrect, then a small insulated tank would be added. Finally, the centerbody of each ramjet carried a tank with 1100 lb; this fuel would be the first to be burned.

The cockpit was another place where drastic actions were taken to save weight. It was to be smaller than that of the U-2, on the grounds that the pilot would only be in it for about one-fifth as long as on a U-2 mission. Many instruments indicating ranges of values would be deleted and replaced with simple warning lights. The vertical speed indicator would be unnecessary as it was of little use during climb and descent, when it would be indicating large values, and the aircraft was expected to spend little time holding during instrument flight. The actual flight controls would be redesigned for minimum weight. Finally, an ejection seat would only be provided on training and ferry flights; for operational flights, a lighter fixed seat would be used.

The environmental system could be kept simple because the pilot would be in a full pressure suit, rather than the partial pressure suit worn at that time by U-2 pilots. The cockpit would be pressurized with nitrogen, rather than air. Boiling off water from a radiator built into the skin of the aircraft would cool the pilot and equipment bay.

The landing gear would technically be tricycle gear, but the main gear would be mounted on a single strut. This was apparently done to save weight on a second main gear strut. The result was that the aircraft could tip to the side at low speeds, and that U-2-style pogos under the ramjets would be needed on takeoff. To save weight on brake drums, they would be made lighter that normal. This meant that if a takeoff were aborted at high speed, the brakes would not be able to stop the aircraft in a reasonable distance. Either the pilot would lock the brakes, blow the tires, and come to a halt on the wheel rims, or a barrier cable would have to be used.

Structurally, the A-3 was unconventional. Although normally the wing skin is maintained through the fuselage, here the wing beams (spars) would carry through, and part of the fuselage would be built as part of the wing. This seemed to be the lightest possible structure. Although the primary design had no horizontal stabilizer, there was a chance that further stability analysis would show that one was in fact necessary. As a hedge against that, work on a version with one continued in parallel with the presentation to the customer.

Data from Marquardt and Pratt and Whitney indicated that the ramjets would provide specific fuel consumption at least as good as the turbojets at speeds above Mach 3.0. However, because the ramjets would not be operating in the climb, they would create drag. The proposed solution was to cover the end of each one with a pressurized sack made of Mylar. The ramjets would be ignited at 25,000 ft to help with the climb. Below Mach 1.6, a normal movable spike was not expected to yield optimum thrust. A possible solution that had been tested on earlier ramjets was to place a hood over the inlet to reduce the area. The hood could be ejected in flight after reaching higher speeds.

The use of borane fuel (HEF-3) in the ramjets was expected to give 15 to 30% greater range than JP-150 (a high flash point version of JP-4) for a given weight of fuel. It could provide a radius of operation of up to 2020 n miles. Johnson also considered using JP-150 and refueling from a U-2, but this would have to be done outside denied airspace. That meant that the A-3 would have farther to fly after exiting, and the overall mission radius was reduced. An alternative was "buddy" refueling, in which a tanker A-3 would fly with the mission aircraft and refuel it as it climbed and entered denied airspace, and then turn back before the mission aircraft reached the target. This would give a radius of 2050 n miles. However, this was more complicated, and using HEF was considered to be the best bet. The downside would be the need for special handling of this toxic fuel and possibly a higher RCS of the exhaust plume.

L. D. MacDonald's team did RCS measurements on 1/20th- and 1/40th-scale models. The maximum returns occurred directly from the side. Average returns were the same as for the U-2, with peak values at some angles that were larger. These were much larger than for the all-metal GUSTO 2S, which in turn was larger than for the GUSTO 2A, which incorporated RAM. The largest reflections on the A-3 came from the ramjets and the fuselage. Shielding of the turbojets was considered to be good [77].

Chapter 9

NOVEMBER 1958 LAND PANEL REVIEW

In preparation for the November 1958 review of Lockheed's and Convair's work by the Land Panel, Gene Kiefer laid out three issues that he felt had to be addressed in the review. The first was the relative detectability by radar of the two designs. Frank Rodgers's team had given both companies advice and had tested their models. To get an idea of how much room there might be for improvement, Kiefer suggested to Rodgers that he get the best estimates of RCS from Lockheed and Convair, analyze them, and report on the results to the panel.

Second, Kiefer wanted performance numbers and the resulting target coverage. He realized that the panel would have to trust the numbers presented to them, although the Agency (CIA) could confirm the target coverage if given the ranges.

Third, he wanted information on ramjets—the expected complexity, reliability, and development times for the engines for the two companies. He proposed to get information from both Pratt and Whitney and Marquardt by either interviewing Perry Pratt and Ray Marquardt before the meeting, or by having them attend and answer the panel's questions.

Finally, Keifer mentioned that he had discussed with Johnson the results of radar measurements of an F-101 burning HEF. The initial results were discouraging, and Kiefer asked Leo Geary to get more complete information. Johnson reported that HEF tests in the ramjet of an X-7 gave disappointing performance figures [78].

CHOOSING FISH

On 12 November, both Johnson and Widmer returned to Cambridge to present their complete designs to the Land Panel. Johnson showed the A-3, and Widmer showed a completed FISH design. Separately from those presentations, the panel had also been reviewing reconnaissance aircraft proposals made by various manufacturers to the military as well as aircraft already under development by the military. Three days later the five principal members of the panel (Land, Perkins, Purcell, Donovan, and Stever) summarized their findings in a memo to James Killian, the President's science advisor:

> We recommend that the development of a new aircraft be undertaken at once on a highly expedited and sensitive basis in order to retain our ability to conduct special reconnaissance. We recommend further that the former proposal utilizing the B-58 to launch a newly designed reconnaissance vehicle be selected for this purpose since this aircraft appears to best meet all of the desired technical features. Although a more detailed comparison may reveal that the latter proposal could be developed at somewhat less cost, this unstaged system does not appear able to meet all of the desired technical features with the same success. In case the system we recommend is not acceptable, we would wish to review other alternatives before recommending firmly a second choice [79].

The second choice was, of course, the A-3. They deemed it "technically somewhat less desirable" because of its lower speed and a range only $\frac{3}{4}$ of the target when it used conventional fuel [80]. They saw that it could only get the desired range by refueling at supersonic speed or by using HEF, with its handling problems. It was also "more susceptible to detection and tracking by radar" and would have a reduced payload [79]. The panel recognized that ramjet development could be the limiting factor in getting either FISH or the A-3 operational.

On Wednesday, 26 November, Bissell called Johnson with the bad news. He said that Gene Kiefer had spoken with two of the panel members to get their own views. Although the issues had been complex, the choice between the Lockheed and Convair designs had been mostly even—except for the RCS measurements, where Convair was clearly ahead. However, Bissell said not to regard the issue as closed. Moreover, he said that the "users" felt that the Lockheed scheme of a single aircraft was "immensely better."

Bissell and Johnson discussed what would happen next. Frank Rodgers would visit Burbank for further discussions on antiradar techniques. Within two weeks a program for development would be started on a basis of "maximum urgency," with a final decision to be made in three months. During that time, Lockheed should study the use of the "iron maiden" design. And Johnson was to submit a budget for the three months work, which was not to include any manufacturing. Johnson also spoke with Kiefer, who said that they could use other engines.

MORE FUNDING

Until now, both Lockheed and Convair had been covering the cost of their work using internal research and development (IRAD) funds, with Convair also receiving some CIA support. With the recommendation by the Land Panel that the development of the U-2 follow-on aircraft be pursued, the CIA and the Bureau of the Budget signed a memorandum of understanding about the government's funding the work. The memo anticipated that the design and testing to determine the final configuration of the aircraft would take four to six months at a cost of $4 to $5 million, which would be made available from the Agency Reserve upon approval of the work.

It was obvious that the total cost of the system could not be estimated accurately, but that it would be at least $100 million and would probably be higher. The agencies agreed to hide the money. If full development were approved by mid-1959, then $75 million would be made available from the Department of Defense budget for fiscal years 1959 and/or 1960; the money would not come from either the CIA budget or Agency Reserve [80].

White House Approval

By mid-December, Killian had briefed the President. At 0900 hrs on 16 December, Eisenhower met with his Board of Consultants on Foreign Intelligence Activities. Eisenhower questioned continuing the overflight program. He recalled how enthusiastic the group had been for the U-2, but that in operation it had been tracked on almost every mission flown west of the Ural Mountains. He pointed out that the successor would have even higher performance.

Eisenhower's main concern was whether the intelligence gathered was worth the resulting tensions between the Soviet Union and the United States. He felt that although they had located enough targets (should the United States have to attack) the U-2 did not solve the most important problem, detecting a Soviet surprise attack. After discussion, the Board felt that the intelligence gained was nevertheless "highly worthwhile" [81]. And so the U-2 overflights and development of the follow-on continued.

Second Thoughts

Perhaps because Bissell's team was concerned about the Land Panel's choice of FISH, the Development Projects Division (DPD) operations group undertook their own comparison of FISH and the A-3. After seven months of operating the U-2, they had a wealth of experience to draw upon, in every area from aircraft operations and overseas basing to Presidential permission and Soviet responses. DPD also went to the Air Force's Rome Air Development Center's Factor Analysis Board for an independent comparison of the vehicles. In preparation, the study team wrote two documents, one outlining the operational requirements and another giving specific selection criteria.

REQUIREMENTS

The operational requirements document covered the general areas of survival, operational utilization, and quality of intelligence provided. Survival of the vehicle was deemed the highest priority. The team was hopeful that the vehicle would not be detected at all, but recognized that Russian radar and infrared detection capabilities and anti-aircraft weapons would improve. If the vehicle had the highest possible performance, its usable lifetime would be increased. Survivability would also be improved through clever tactics, such as choosing a flight path that would exploit dead zones in radar coverage, reduce the time exposed to radars that could not be avoided, and confuse specific radars (such as arriving when other aircraft would be present and causing blips on radar scopes).

The analysis of operational utilization included ways to minimize the chance of exposure of the project and the details of employment of the vehicle. Reducing the risk of exposure would require having a minimum number of people involved in operations, operating the vehicle from a "ZI" base (zone of the interior, i.e., the continental United States) or from an aircraft carrier, and confusing the Soviets as to the flight path and ownership of the vehicle.

Operational employment covered a variety of factors. For timely operations, the vehicle should be able to fly a mission after a 24-hour warning, fly another mission within three to four hours of landing, and after disassembly fit into an existing cargo aircraft for evacuation. Adequate payload capability was desired to allow adding new intelligence gathering systems, including ELINT and high-resolution radar. The U-2's navigation systems were considered to be inadequate for the follow-on vehicle, which would be flying much faster and which might be flying over the north magnetic pole, rendering compasses unreliable. The new aircraft would have to minimize crew fatigue by being comfortable and simple for a single pilot to operate. A capsule escape system was thought to be safest, whereas a full pressure suit with a tumble-free ejection seat would be the minimum acceptable. Other considerations included operation in 1961 and improved performance in 1963–1965.

The requirements document also described the expected Soviet process of attempting to detect and intercept the vehicle, as well as the resulting political reaction. Finally, it outlined the types of intelligence needed, primarily indications of imminent hostilities.

COMPARISON

The final study, released on 15 January 1959, compared the two aircraft in 13 major areas. FISH was deemed better in terms of speed (Mach 4.0 vs the A-3's 3.2), radar immunity (because the A-3 was deemed to have no RCS reduction features), infrared reduction, navigation (inertial navigation vs the A-3's use of a system similar to the U-2's), fatigue, escape, and payload (560 lb vs the A-3's 215 lb). The A-3 was ahead in turnaround time (3–4 hours vs 8–12 for FISH), sortie rate (two per day vs one for FISH), and maintenance (twice that of the U-2 vs FISH's four times that of the U-3). FISH might also have had a longer range (4150 n miles and 5150 with refueling), but the Agency has oddly redacted the A-3's range from the released study.

The study concluded,

> As pertains to the desired criteria, both vehicles are deficient. The super-hustler [sic] comes closer to meeting the criteria than the A-3. Since the A-3 is deficient in some of the most important areas such as: range, employment, radar immunity, navigation systems, fatigue, escape, and pay load areas, it is the opinion of operations that the super-hustler is superior to the A-3 [82].

In other words, the study confirmed the finding of the Land Panel that FISH was a better choice than the A-3.

FURTHER STUDIES

On 22 December, Convair was given the go-ahead to proceed with development of FISH. One week later manufacturing work began in the Convair model shop and in the Fort Worth nuclear manufacturing building. The work focused on four areas: wind-tunnel modeling, materials research, construction of a full-scale "electronic" (RCS) model, and construction of a full-scale inlet duct assembly for use by Marquardt, the engine contractor.

The model shop built four "force" models and four inlet models for testing in off-site wind tunnels. Materials research covered design, manufacturing, and tooling with materials that included PH15-7MO, Iconel-X, Rene-41 steel, plastics, and glass. Eventually a number of major components were fabricated, including a wing-box section, a leading-edge section, an acoustic panel, and a fuel-tight section.

The RCS model and a rotator were fabricated in Fort Worth and transported to test ranges in a specially built trailer. As studies progressed, the model was continuously modified and once was repaired following a accident. Originally two ducts were to be built and tested at pressure of 70 psi and temperatures of up to 1400 deg. Eventually, only a single duct, using Iconel-X, reached the stage of welding in final assembly [83].

White Convair began work, Bissell's staff proceeded with their own planning for its production and operation. On 26 February, representatives of the operations, research and design (R&D), material, administration, security, personnel, contracts, and comptroller offices met to answer questions about the logistical and contractual support for the vehicle. The first question was how many FISH articles to build. The minimum was agreed to be seven—two or three for flight test and four for operations. However, they knew all too well that they should expect to lose 30 to 40% of the aircraft during the first year, and so the recommended number to buy was 12. That would have the advantage of providing more aircraft for pilot checkout and allowing operations to continue when aircraft were out of service.

They also discussed the number of B-58 carrier aircraft to obtain. The minimum was one B-58 for every four FISH. The optimum would be one B-58 per FISH, so that the same aircraft and crews could always be paired, a team training and operational concept. Gene Kiefer was tasked with discussing the number with Bissell and with touching base with the person preparing the "treaty" with the Air Force to obtain B-58s.

The next question was how long would the test and initial operational period be. The opinion of the staff was that testing would take one year and operations 18 months, although there would likely be some overlap.

There was a long discussion of where FISH would be based. The consensus was that Carswell Air Force Base, which was near Convair's Fort Worth plant, would be the best location. Because Convair had a "special area" at Carswell, security would be easy to maintain. The Gulf of Mexico was envisioned as the test area because of the lack of air traffic and the absence of people to complain about sonic booms. No plans were made for foreign basing, although it was expected that flights could be staged through Alaska and Greenland. Postflight retrieval could be done anywhere in the world, especially if FISH were to land on an aircraft carrier. The security and operations staffs would study the security situation at Carswell and brief Bissell, and the staff would work with Convair to concoct a cover story for what would be a strange-looking aircraft.

The group decided that the Agency should assume responsibility for maintenance of the B-58 carrier aircraft. Presumably this meant that Convair personnel would actually perform the maintenance. Normal B-58 spare parts would be procured from the Air Force, and parts peculiar to the carrier version of the B-58 would come from Convair. With three different engine types in the program—J79s for the B-58s and ramjets and a small turbojet for each FISH—engines were seen as a potential problem. They would have 150% spares for the first four FISH, when more problems were expected, and 50% for the later aircraft.

One of the lessons of the U-2 program had been that technical manuals supplied by the contractor had been written for maintenance personnel with the skill level typical of the contractor, which was much higher than that of an Air Force technician. However, because they saw little chance of an Air Force version of FISH, this was not expected to be a problem. Testing of the aircraft was discussed. Gene Kiefer was tasked with telling Convair that they should plan to build a static test article and test it to destruction, as well as telling them that they would be responsible for all systems testing, which was the way U-2 testing had been carried out.

Finally, the group considered whether the Agency should have an individual stationed at Convair for liaison. (No action was taken on this, but it turned out that this was eventually necessary at Lockheed during the development of the A-12 [84].)

On a related front, support aircraft would be needed for the testing and operational phases of FISH. Because Air Force personnel performing maintenance and other operational functions on the conventional aircraft at the U-2 detachment at Wiesbaden would soon be coming under direct Agency control, Bissell felt that it would be reasonable to use these resources when necessary. The subject had to be broached with the Air Force, so he asked Kiefer to "... approach this matter somewhat cautiously with Leo [Geary] ..." [85].

On 20 March, Widmer conducted a review of FISH's propulsion system. Among others, Savage and Matteson of Convair and Gene Kiefer from the Agency attended. One conclusion was that the first year of flight test would see about 200 hrs, some of which would be captive and some of which would be free. Each aircraft would get about one hour of flight per week. The production rate would be one aircraft every other month. It was expected that a ramjet could be used for about 6.9 hrs before being overhauled. Nine aircraft would need two engines each, and 200% spares would require a total of 54 engines. Widmer was concerned that the expected high specific fuel consumption (SFC) during climb might not be acceptable [86].

BACKCHANNEL

Leo Geary felt that the selection of FISH was wrong. In a bid to change the choice—and with Bissell's knowledge—he went to General Thomas D. "Tommy" White, the Air Force chief of staff. He argued that Lockheed had designed the better aircraft, one that the Air Force could use. He urged White to fund development of the Lockheed design if the CIA chose not to (Telephone interviews with Leo P. Geary, 21 Aug. 2002 and 24 May 2003).

Chapter 10

LOCKHEED STEALTHY DESIGNS

The effect on Johnson of the rejection of the A-3 was to make him focus on reducing the radar cross section of his next designs, while sticking with a vehicle that could launch itself (unlike Convair's FISH). Over the next two months his team explored the A-4, A-5, and A-6 designs. The general approach was to have a small physical size, to hide vertical surfaces above the wings, and to blend the wings and fuselage. He would explore combinations of turbojets, ramjets, and even rockets for propulsion [87]. The designs incorporated variations of Frank Rodgers's "iron maiden" shape, having chines that began at or near the nose and merged into the wing leading edges.

THE A-4

On 26 November, the same day he spoke with Bissell, Johnson began laying out the way forward with the A-4. He wrote down the ground rules: 1) basic *nonrefueled* range on JP-150 is 4000 miles, 2) do not use JT-12As, 3) radar—cross section vitally important, 4) cruise alt. can be reduced some (95 to 91,000—target), 5) basic self-contained system, 6) basic cruise Mach No.—3.2, 7) no honeycomb, and 8) alright to use rocket assist. Johnson's first sketch placed two afterburning J57-43A turbojets and two 52-in. ramjets in the fuselage (Fig. 64). The turbojets were fed air by an inlet on either side of the fuselage behind the equipment bay and the ramjets by inlets below the fuselage, similar to FISH. The vertical stabilizer was blended around the engines into the wings, giving a fat appearance and eliminating a flat surface that would have reflected radar energy back at the transmitter while in level flight. The 15-deg slope on the side would allow the aircraft to bank up to 15 deg for turns before giving a direct reflection.

The design would have had an empty weight of 35,200 lb. Weighing 72,000 lb at takeoff, by the time it had reached cruise altitude, it would have burned 12,000 lb of fuel. That 60,000-lb weight would have given it a range at altitude of 2700 miles and a total range of 3000 miles if a 300-mile climb phase were included. Johnson decided that the ramjets would have had to be increased to 58-in. diam to provide the 12,400 lb of thrust needed to maintain cruise altitude.

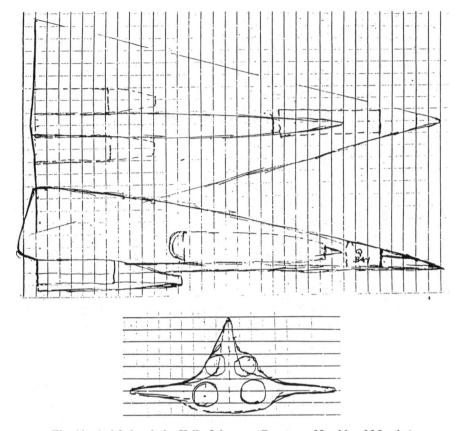

Fig. 64   A-4-2 sketch, by Kelly Johnson. (Courtesy of Lockheed Martin.)

A week later, on 3 December, Johnson began an alternate design (see Fig. 65).

> Consider a revision to A-4. Can we use 1 ramjet of large size and two turbojets? The airplane would be about as large as indicated previously (35,000# empty) but might lay out better. Also, can we use JT-12's with A.B. with <u>no</u> costly development problem to M = 2.0? (Johnson, C. L., Archangel project design notebook, entry for 3 Dec. 1958, pg. 7.)

Johnson's computations indicated that 9000 lb of thrust would be needed; that would require a ramjet diameter of 63 in. To improve takeoff acceleration and initial climb performance, he added an Aerojet "super performance rocket," which would deliver 10,000 lb of thrust, burning a mixture of hydrogen peroxide (H_2O_2) and JP-5. Two minutes of thrust would be delivered by 4880 pounds of H_2O_2; it would be carried in wing-tip drop tanks, which could then be discarded, eliminating their weight for most of the duration of the mission.

LOCKHEED STEALTHY DESIGNS

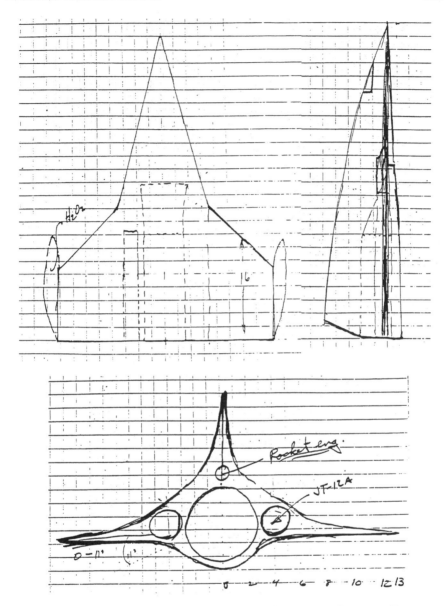

Fig. 65   Second A-4 (later renamed A-5) sketch, by Kelly Johnson. (Courtesy of Lockheed Martin.)

The next day, Johnson checked the rocket engine data with Perry Pratt and Bill Gore, at Pratt and Whitney. They told him that the rocket could be throttled from 3000 to 10,000 lb. It could run for two to five minutes on 385 gal of fuel that was mixed at 90% H_2O_2 to 10% JP-5, although in principle there would be no limit on the operating time if more fuel were available. The engine was expected to begin testing in the summer of 1959.

He then called Bill Sens to check on the JT-12A turbojet. Johnson wanted to know whether it would be practical to shut it down during the Mach 3 parts of the flight. Sens told him that the big problem would be heating of the engine; its compartment would have to be kept at 300 deg or less, and that some airflow and circulation of fuel would be needed. The maximum speed at which the JT-12A could operate would be Mach 2.0 to 2.1. Two years would be needed to get a production version, and the development cost would be 10 to 15 million dollars.

Johnson gave the design to Dan Zuck to produce a general arrangement (GA) drawing; it was renamed the A-5 because it was significantly different from the A-4 concept.

On 9 December, Johnson again revised the A-4 design, using a J58 instead of the JT-12As. He concluded that this had several advantages: 1) has a lighter overall weight, 2) runs throughout flight and solves a difficult accessory problem, 3) has overall better performance on turbojet only, 4) is designed for 3.2 now, 5) has smaller ramjets, and 6) has three-engines vs one-engine performance at altitude.

The accessory problem solved by the J58 was that the JT-12A turbojets would not be available to provide power for aircraft systems while at cruise speed. The only disadvantage would be that on landing there would only be a single engine running, the J58. Working through the numbers and comparing this design with the previous one, he concluded that the two smaller ramjets would have a smaller specific fuel consumption. With an 850-ft^2 wing area, the JT-12A version would weight 16,000 lb at the end of cruise, while the J58 version would weigh 18,500 lb. He wondered "Which is best airplane?"

A zero-velocity ejection seat would reduce the risk of landing with only one engine. Johnson concluded that the J58 version with two ramjets would probable be superior, based on "... overall safety, performance, simplicity, & accessory drive" (Johnson, C. L., Archangel project design notebook, Lockheed ADP, Burbank, CA, entry for 9 Dec. 1958, pg. 17).

Johnson worked out two variants of the A-4. The A-4-1 had no ramjets and a single J58. Its predicted operating radius was 1780 n miles, with a midrange altitude of 89,000 ft. The A-4-2 included a 34-in. ramjet on each wing tip in an attempt to increase the altitude to 92,000 ft. However, this reduced the radius to 1320 n miles, just as it had when ramjets were added to Archangel I. The lesson was that simply adding ramjets and reallocating some of the existing fuel volume for their use might improve altitude, but it would be at the cost of range.

Johnson assigned the detailed work to Baldwin, who completed the GA drawing of the A-4-2 in two days (see Fig. 66). The chines began just forward of the cockpit and ran back to the wing leading edges. Perhaps the most unusual-looking feature was the very long and thick vertical stabilizer, which began just behind the equipment bay and ran 38 ft back to the end of the aircraft.

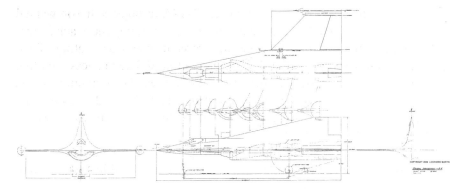

Fig. 66   A-4-2 three-view drawing, by Ed Baldwin. (Courtesy of Lockheed Martin.)

This first attempt at making a small airplane stealthy was not promising. The A-4-1 was short of the range target by over 200 miles and short of the altitude target by 1000 ft. Trying to improve the altitude only hurt the range.

THE A-5

The second A-4 concept, with one ramjet and two JT-12A turbojets, had become the basis for the A-5. In early December, Baldwin did the calculations for replacing the A-3's two 40-in.-diam ramjets using circular inlets with a single ramjet using a half-circle inlet. He came up with an inlet radius of 48.4 in. and a circular outlet with a diameter of 82 in., almost 7 ft. He gave the information to Dan Zuck to use in the A-5 design, which Zuck finished on 5 December (Fig. 67).

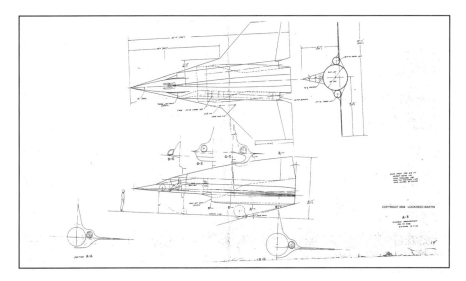

Fig. 67   A-5 general arrangement drawing, by Ed Zuck. (Courtesy of Lockheed Martin.)

The enormous ramjet occupied the center of the aft part of the fuselage, with its inlet under the fuselage and just behind the equipment bay. The JT-12As were on either side of the ramjet, with half-cone inlets on either side of the cockpit and above the chines. On top of the ramjet was the Aerojet rocket engine burning hydrogen peroxide. Thin wings were flush with the bottoms of the turbojets, and fillets blended the wings around the turbojets to the ramjet. As with the A-4, the vertical stabilizer was fat on the bottom to provide 15-deg sloping sides. It covered the rocket engine and ran forward over the top of the canopy; the forward section of the stabilizer would move up when the canopy was raised.

On 29 December, Johnson sat down with Dick Boehme, and they outlined the projected engineering costs for the A-5. They assumed the work would require 40 engineers for three months. At 175 hours per month, a cost of $11 per hour, and a 9% profit, the total would be $252,000. L. D. MacDonald would provide 12 men working for half of that time, for an additional $38,000. The total cost would be $290,000.

With its combination of turbojets for takeoff and landing, a rocket for boost to altitude, and a ramjet for cruise, the A-5 had the most complex propulsion system of any of the designs Lockheed explored in the course of Project GUSTO. It was also the smallest aircraft in the series.

The A-5 was able to cruise at Mach 3.2 at 90,000 ft, but had an operational radius of only 1557 n miles, midway between that of the A-4-1 and A-4-2. It suffered from the problem of not having a means to power aircraft systems during the ramjet-only-powered cruise.

THE A-6

During December 1958 and January 1959, Dan Zuck worked through at least nine variants of the A-6 design, which had features later used in the final A-12 design. These included fuselage chines and twin vertical stabilizers canted inward. Like the A-5, all were propelled by an afterburning J58 turbojet and two 34-in. ramjets, but without the rocket booster. The J58 was placed in the main fuselage and fed air from a semicircular inlet under the cockpit through a bifurcated duct that ran around the equipment bay behind the cockpit. The ramjets blended into the bottom of the wing-fuselage junction and fed from two inlets that were approximately 120 deg of a circle and which were placed almost halfway back from the leading edge. The wings were about 1100 ft^2, and each vertical was a bit over 200 ft^2.

In an effort to save weight during climb and cruise, the A-6-5 (Fig. 68) featured full-size landing gear that would support the weight of a fully loaded aircraft and that would be dropped after takeoff. For landing, when the aircraft weight was at its minimum, three much smaller gear would be extended; this would give a ground clearance of about 5 in. below the ramjet inlet. Although

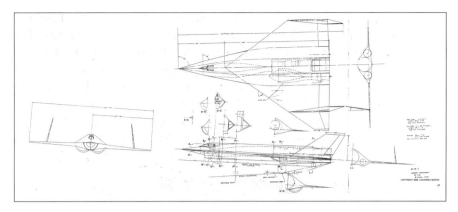

Fig. 68 A-6-5 general arrangement drawing, by Ed Zuck. (Courtesy of Lockheed Martin.)

the A-6-5 could cruise at Mach 3.2 and 90,000 ft, its radius was only 1287 n miles, even less than that of the A-4-2.

The A-6-6 had a third vertical stabilizer on its dorsal (back) surface (Fig. 69). To reduce its radar return, Zuck borrowed a feature from the A-4 and A-5 and gave it a very wide base, which allowed the sides to have a 15-deg slope inward. This matched the slope of the two outer verticals. He discarded the droppable gear concept and used tricycle gear.

On the A-6-9, Zuck moved the J58's inlet back behind the equipment bay and extended the aircraft's length to 68 ft. He also moved the twin verticals inward and tipped the outer section of the wing (beyond the verticals) up 25 deg (Fig. 70). The trailing edge of the rudders swept forward, rather than sweeping back beyond the trailing edge of the wing. This forward sweep helped to hide the verticals from radar, a technique that was eventually adopted in the final A-12 design (Fig. 71).

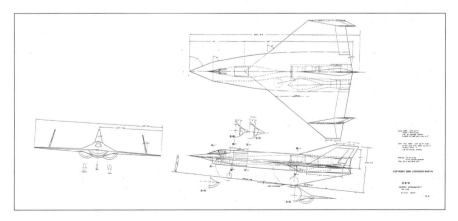

Fig. 69 A-6-6 general arrangement drawing, by Ed Zuck. (Courtesy of Lockheed Martin.)

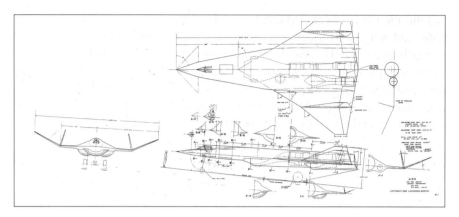

Fig. 70   A-6-9 general arrangement drawing, by Ed Zuck. (Courtesy of Lockheed Martin.)

One of the major challenges of the A-4 through A-6 designs was fitting all of the systems into the available space, especially getting enough fuel into the vehicles. Baldwin estimated that in a thick wing 94 to 95% of the volume can be filled with fuel. In trying to fill every possible space with fuel, they looked for places to drill holes to let fuel flow through structural members without affecting their strength.

Working against them was the need to add extra structure to blend the wings, body, and tail to achieve a stealthy shape. In the end, the fuel fraction of the designs was too small, and none of them could achieve the required operational radius of 2000 n miles.

Fig. 71   The A-6-9 is shown in comparison with an early A-12 design. (Courtesy of Lockheed Martin.)

Johnson finally concluded that the goals of achieving maximum speed, altitude, and range performance as well as minimum radar cross section were mutually exclusive. He also decided that ramjet technology was too immature to be used for long range cruise. The small, stealthy designs seemed to be a dead end. Bissell's strategy of using Convair to force Johnson to focus on stealth had worked, resulting in a deeper understanding of techniques to achieve reductions in RCS and how they affected aircraft performance. The problem was that it had not resulted in a design that met all of the requirements.

ARROW: LOCKHEED'S FISH

Evidently because the Land Panel had perceived FISH as a radical, high-risk approach, in the November meeting they asked Johnson to perform a sanity check on the design. The task was to design a ramjet-powered vehicle that could be air-launched from under a B-58 at a speed of Mach 2.0 and an altitude of 45,000 ft and that could cruise above 90,000 ft at Mach 4. If the Lockheed design approximately matched the performance and RCS of FISH, then the work of Widmer's team would be validated.

Johnson assigned the work to Zuck and Combs. They came up with two similar designs, named "Arrow I" and "B-58 launched vehicle" (see Figs. 72 and 73). Both used a pair of 40-in. ramjets for cruise and one JT-12 turbojet for landing. Like FISH, they could cruise at Mach 4 above 95,000 ft. One had a radius of operation of 2208 n miles and the other 1736 n miles, which bracketed FISH's expected radius of 2075 n miles.

Lockheed also performed RCS measurements on these designs (see Fig. 74). Because the Lockheed designs used straight leading and trailing

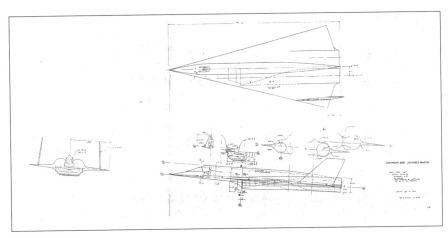

Fig. 72   Arrow I general arrangement drawing, by Dan Zuck. (Courtesy of Lockheed Martin.)

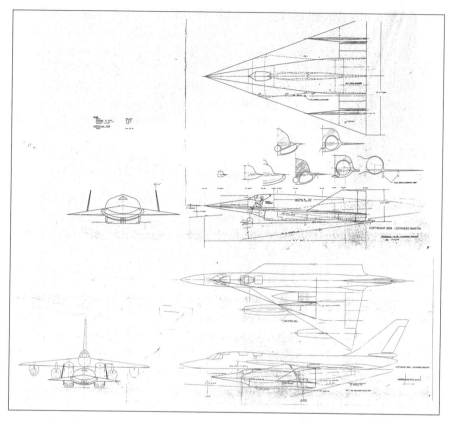

Fig. 73 "B-58-Launched Vehicle" general arrangement drawing, by Henry Combs. Note wing openings for B-58 main landing gear. (Courtesy of Lockheed Martin.)

wing edges, unlike the curved edges on FISH, the radar reflection patterns undoubtedly differed.

Lockheed's designs validated the Convair performance numbers. However, they found the same operational problems that Convair was already wrestling with. The reliability of the ramjets in the cruise phase was in doubt; the JT-12 turbojet only provided enough thrust to flatten the landing approach and not enough to go around and try the landing again; the clearance between the parasite and the B-58 and between the parasite and the ground was minimal; and there was no way for the pilot to eject while the vehicles were mated [87].

NONSTEALTHY DESIGNS

At the beginning of January 1959, before Zuck had completed the A-6 design studies, Johnson launched his team on a series of small but nonstealthy designs. He set three objectives:

1) Design a maximum performance aircraft consistent with an 18- to 24-month development schedule.

Fig. 74 Arrow I RCS model is shown suspended in the anechoic chamber. (Courtesy of Lockheed Martin.)

2) Make no performance concessions to reduce radar cross section. Use a similar configuration to A-1 and A-2, but with smaller dimensions.

3) Use one J58 afterburning turbojet plus two 34-in.-diam XPJ-59 ramjets burning JP-150 only, not HEF [86].

There were three major designs in this approach: the A-7, the A-8, and the A-9.

THE A-7: UGLIEST AIRPLANES IN THE WORLD

Like Johnson, Ed Baldwin believed that airplanes should look good in addition to performing well. In the first half of January, he produced three variants of the A-7 design, which he later called "the ugliest airplanes in the world." The A-7-1 and -2 were about the same size, both having a 767-ft^2 wing and lengths of 80.83 and 73.33 ft, respectively. Each was powered by a single J58 and two ramjets. They differed in their wing and inlet arrangements.

The A-7-1 had a high wing and its J58 had a large inlet with a boundary layer bleed on the bottom of the fuselage: "... that one was bad enough so we made two other versions" (Baldwin, E. P., oral history, unpublished audio recording). The A-7-2 had a low wing with a half-cone inlet on each side of the fuselage and was "very ugly. You do what you have to do or what seemed like the thing to do at the time" (see Fig. 75).

Baldwin took tremendous pride in his work, but was able to criticize it.

The A-7-3 looked like a larger A-7-1, with a wing area of 990 ft^2 and a length of 93.75 ft (Fig. 76).

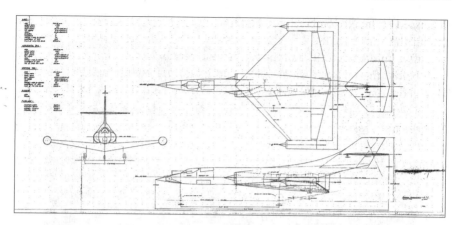

Fig. 75 A-7-2 general arrangement drawing, by Ed Baldwin. (Courtesy of the family of Edward P. Baldwin.)

No drawings of the A-8 and A-9 are available. However, they were similar to the A-7 in that they used a J58 turbojet and two ramjets. The designs were probably drawn by Herb Nystrom.

None of the A-7 through A-9 designs were judged to be workable. Their mission radius of typically 1637 n miles was too small, and the midmission altitude of 91,500 ft was far below that of Archangel II, the last design that had made no concessions to reduced RCS. So a small airplane—even a nonstealthy one—was out.

SUPERSONIC REFUELING

After seeing the A-7 designs in January, Bissell's office did a study of the feasibility of in-flight refueling at supersonic speeds, which had been

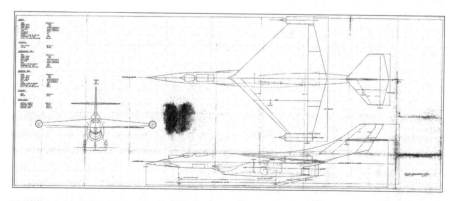

Fig. 76 The A-7-3 was a high-wing version of the A-7-1. (Courtesy of the family of Edward P. Baldwin.)

discussed at the November Land Panel review. The hope was to find a way to extend the range of the Lockheed designs. They decided that the receiver aircraft would have to fly far enough behind and below the tanker to keep its nose below the tanker's shock wave and its tail below the tanker's exhaust. This geometry meant that the refueling boom would have to be 85 ft long, would weigh 12,000 lb, and would have half the drag of the basic airplane. This did not look promising.

On 3 February 1959, an Air Force major on Bissell's staff met with Kelly Johnson in Burbank to discuss the concept. Johnson's off-the-cuff evaluation was that although the boom was much too long, refueling in general would be possible if the receiver aircraft could descend to 80,000 ft, where the air density would permit the necessary fine control for formation flying. He suggested that a smaller boom would work if the two aircraft flew side by side and the boom extended laterally from the tanker's wing. The major's memo said that this could not be done with the A-7-3 because of the ramjet on the wing tips. This suggests that they were considering an A-7 variant with no ramjets. The biggest problem would be the inability of the pilot to see the boom, which would be plugging into his aircraft well behind the cockpit.

Johnson then explained why the concept would be impractical. Because the Agency did not want the refueling to take place over denied territory, and the altitude would be 80,000 ft, refueling would only be able to replace the fuel expended during takeoff and the climb to 80,000. With full tanks, the receiver would then have to burn off some fuel to be light enough to reach the penetration altitude of 90,000 ft. If a penetration altitude of 80,000 ft were instead acceptable, then the aircraft could be designed to take off with enough fuel to reach the desired range of 4000 n miles, and the whole complexity of refueling could be avoided [88].

Although supersonic refueling has apparently never been accomplished, the concept did not die completely. Years later A-12s were flown in formation at high speed in order to judge whether there was sufficient stability and control to refuel. The shock wave from the nose of the lead aircraft interfered with the trail aircraft and made stable flight impossible (Interview with Frank Murray, Las Vegas, NV, 1 Oct. 2007).

EISENHOWER MEETS THE FLYING SAUCER

On 10 February 1959, the President met with Killian, Land, and Purcell to discuss progress on reconnaissance projects. One of these was the Corona spy satellite project, where the cameras were progressing well, but the Thor launch vehicle and the recovery operations were still problematic. They also discussed the discovery of a layer of the atmosphere at about 70,000 ft, which should act as a "sound duct" to conduct the sound of Soviet missile launches over long distances. By placing at least six balloons with listening devices in the "duct," it was thought possible to triangulate the launches. Eisenhower discussed how certain senators, egged on by defense contractors pursuing contracts, were leaking classified information and painting a pessimistic picture of the American security situation.

Killian then reported on the progress on the new reconnaissance aircraft. Purcell described how its "low radar visibility" was caused by its shape. He said that the best shape would be a flying saucer and the next best a triangle with a smooth underside. Eisenhower recognized that with high speed in thin air it should have a long range. Purcell described how it would move so far between radar sweeps that it would be very difficult to track—the blip-scan theory. Land added that the payload would be 700 lb, and the vehicle probably would not be seen [89].

Two days later Eisenhower again fended off a request for more U-2 overflights. In a meeting with Eisenhower, Quarles, and Twining, McElroy described how he had been able to blunt Congressional criticism of the state of American intercontinental ballistic missile (ICBM) work, but that there was concern over the validity of the intelligence estimates. McElroy pointed out that there was no knowledge of any Soviet ICBM sites and proposed a series of overflights to attempt to locate them. He cited the opinion of the Joint Chiefs of Staff that the U-2s would not be shot down, and Twining joined the fray, adding that the Joint Chiefs would like more information.

Eisenhower pushed back, saying that he wanted to minimize overflights until the follow-on aircraft was available. Quarles pointed out that this was not expected for 18 to 24 months, but Eisenhower was not persuaded. He said that he doubted that the Soviets could build many launchers in the next year, and that was based on U.S. construction capability; he noted that the United States had consistently overestimated the ability of the Soviets to outperform them and cited the overestimation of the bomber gap two years earlier. He wanted to avoid any provocations, especially because the confrontation over Berlin was becoming a crisis. He said that nothing would make him request authority to declare war faster than a Soviet intrusion into American airspace. He felt that reconnaissance satellites were the future for this kind of reconnaissance.

In the end, Eisenhower said that he might approve one or two flights. One had already been approved for the north of the Soviet Union and was on hold until March, awaiting improved sun angles. He remarked that the next mission would tell whether the Soviets had developed adequate surface-to-air missiles [90].

Chapter 11

REFINING FISH

After the November 1958 meeting, Convair had also begun a redesign of their aircraft. While Johnson's team worked on the A-4 through A-11, Widmer's team refined FISH and developed plans to start production. A series of eight task change requests document the changes in the work from early February 1959 through early June.

The Agency (CIA) was suddenly faced with the need to clear an enormous number of Convair employees, from engineers to shop personnel to administrators. The CIA Office of Security Support established a temporary office on location that had a staff of eight professional and four secretarial people who rotated in and out. The office also made use of a commercial investigation team to assist them. The peak of the work was between February and August 1959, and by the middle of the year 700 people had been cleared, apparently without any indication that a crash government program was underway [91].

Radar studies of FISH began on 19 December, with a $\frac{1}{8}$-scale model having an "unsoftened" wing. According to Rodgers, these compared well with tests on a $\frac{1}{46}$-scale model. Four days later, experiments began with a softened version of the wing, that is, one with notches filled with RAM. Plans were made to begin testing full-scale inlet and ejector models in early January. At the same time, 70 megacycle measurements would be made of the steel pole that would support the full-scale model to see how big the return from the pole would be. Although the full-scale model had not yet been built, its tests were planned for the last week of January. It was understood that for the full-scale measurements to be meaningful, the model would have to simulate all of the changes in materials and other discontinuities that would exist in the actual aircraft [92].

FACILITIES

At the end of 1958, the work at Convair was formally moved under Project GUSTO. In mid-January, Kiefer and others visited Fort Worth to discuss the scope and organization of the work. They agreed to take a "hard look" at the program for four months, to be sure that Convair could deliver, especially in the areas of security, personnel, facilities, production, and capability for expansion. They realized that at the peak level of effort it would be difficult to stay below the desired number of cleared individuals. There was discussion of showing the Convair workers a film about the U-2 as a way of impressing them with the importance of the work, a suggestion which security did not like.

The facility itself was a source of worries about security. Although Convair had done classified work for the Air Force, GUSTO work had to be segregated and kept out of sight of uncleared personnel. The initial plan was to move the work into a corner of an existing building and tighten physical security. They discussed acquiring additional land and building a new 300 × 300-ft structure to hold the project. The security of using a wind tunnel was also a worry.

Bissell was also concerned about the concept of a parasite aircraft. With his background in statistics, he understood very well that having two vehicles would significantly increase the chances of a failure. In the 20 January DPD staff meeting, he directed DPD Operations to do an independent study of FISH's payload, performance, suitability for overseas bases, and flexibility for recovery, such as onto an aircraft carrier. He suggested making discreet inquiries at Edwards Air Force Base to see whether any Air Force groups there had experience with operation of a "dual vehicle." He also asked for an evaluation of the capability for in-flight refueling. Finally he suggested starting discussions on the camera package for the vehicle; Land, Baker, Yutze, and Lundahl were to be included. Bissell also wanted to let Fairchild, Itek, and Perkin-Elmer know that there might be a competition in the offing [93].

The next day, Bissell approved Letter Contract No. HL-4646 for

> ... the procurement of initial studies, tests and preliminary designs of a high altitude supersonic reconnaissance vehicle to replace the U-2. ... The work is being conducted simultaneously with that being performed by Lockheed on their design version. It is anticipated that upon completion of this portion of the program an evaluation will be held and a decision made to proceed with one of these contractors [94].

Although the Agency was trying to follow the U-2 management model of giving the contractor as much leeway as possible, they still required the contractors to inform them of changes in work that would change the payments. Convair found that they needed to build inlet ducts, a task originally planned for Marquardt. The segment that was needed for engine testing was from the variable geometry location to the engine connection. The first duct was planned for delivery on 30 June and a second on 15 August. This change would be funded by redirecting money from Marquardt to Convair [95].

By late February, Convair had found a need to modify the full-scale radar test model, and the resulting changes to the FISH inlets had to be added to the wind-tunnel model. Convair estimated that the increased costs would amount to $25,000; the Agency hoped that this could be absorbed by underruns

REFINING FISH   145

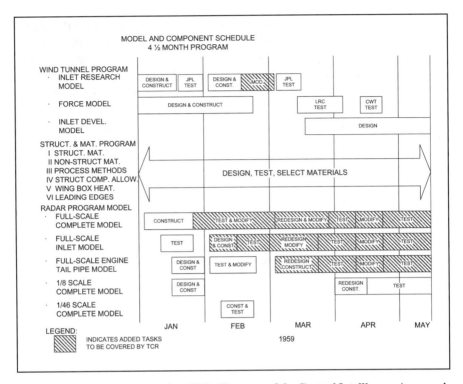

Fig. 77 FISH schedule for spring 1959. (Courtesy of the Central Intelligence Agency via the National Archives and Records Administration.)

elsewhere in they project [96,97]. They issued Task Change Record No. 2 to authorize the work [98]. In a 24 March meeting, Convair informed the Agency that more model modifications and tests were necessary; they were authorized by another change record as an increase in the required work [99].

On 9 and 10 March, Convair delivered a proposed statement of work to Bissell's office. It laid out a four-and-a-half-month design study to produce a design compatible with the requirements that had been laid out in letter CHAM-0085 (Fig. 77). The work would cover four broad areas:

1) *Analysis and design*—This design study will embrace technical analyses, design layouts and tests directed to solutions of electronics effects, L/D, ram recovery, and structural design. Coordination with the engine contractor will be carried out.

2) *Models and components*—Design, construction, and tests of models and components in accordance with Schedule PFY-204-0 will be carried out to provide wind-tunnel data and structural and electronic data.

3) *Manufacturing research*—Preliminary exploration of manufacturing and processing techniques will be accomplished to study feasibility of structural design.

4) *Subsystems*—Studies, preliminary layouts, and analyses of the required subsystems will be carried out including surveys of the state of the art and existing systems developments. This work will serve as a preliminary to selection of possible vendors for the required subsystems [100].

One of the problems was finding a large enough facility to build the FISH vehicles. Neither the main plant nor the experimental hangar had enough room. The Atomic Energy Commission (AEC) building used for the nuclear-powered aircraft project was out because the future of that program was unclear and because it was only half the size needed. There were no other buildings available in the Fort Worth area, and moving production to the Convair San Diego facility presented too many problems.

Convair thus believed that it would be necessary to build a new building. Assuming a go-ahead on FISH production in mid-May, the building would have to be finished by mid-August to prevent a delay in production. To meet a mid-May start, they proposed to start architectural and engineering planning at the beginning of April.

Convair did not want to pay for construction of the building from corporate funds unless they could get a five-year tax writeoff and assume ownership at the end of the five years. They preferred instead to have the contract pay for the building over five years and the government assume ownership [101].

They proposed to rent land for a new assembly building from the Air Force and that the Agency would pay for the building. After five years, the land would revert to the Air Force. Because he felt there were some misunderstandings about the work to be done and because the choice of contractor and even whether GUSTO would continue were uncertain, Bissell disapproved the plan [102]. Instead, he approved $15,000–18,000 to start the architectural and engineering work, with a decision on further work to be made by 1 April [103].

On 2 April 1959, representatives from the Agency visited Convair's Fort Worth facility to estimate the cost of facility modifications and equipment to support FISH production. The report laid out a six-phase building plan and stated that in order to have the first phase (engineering offices) ready by 16 May, construction would have to start in two weeks. Two Agency security officers approved the plans on the spot [104]. One week later, the DPD contracting officer, comptroller, and general counsel signed off on the Definitive Contract HL-4646 with Convair. The contract covered "… the procurement of preliminary studies, tests and designs of a reconnaissance vehicle to replace the U-2," but no facilities construction [105].

By mid-April, Convair and Marquardt had agreed to cancel the 30 June delivery of a ramjet duct to Marquardt; only the 15 August delivery would happen. Because they would only be testing a single duct design, there was a possibility that the design would change before installation in the aircraft [106].

In the first week of May, Convair began a small rearrangement of their plant and continued design work on the new facility. Some engineers were moved and, work was begun on increased security measures. Some equipment like calculators and blueprint machines were added to the budget. The change reflected that costs for some less expensive equipment as well as non-security-related rearrangements would be absorbed by Convair. However, because the original contract scope of work had not covered the security changes, the Agency was billed for the work as an increase in the fixed fee. Finally, Convair noted that the time to order the calculators and blueprint machines for facilities work was approaching [107]. They were ordered a month later [108].

SUBSYSTEMS

In late May, as the next Land Panel review approached, Convair began awarding subcontracts for the design of subsystems for FISH. The first was an air turbine drive, with Marquardt to do a preliminary design of the motor section of the drive and to help Convair prepare a procurement specification by 18 July [109]. While under ramjet power, the system would be driven by 1250°F ram air; at subsonic speeds the turbojet's compressor stage would provide bleed air. The air would drive dual turbines, each of which drove a hydraulic pump and an alternator. The alternators would be developed by Westinghouse under a second subcontract.

The hydraulic fluid and alternator coolant would be cooled in a heat exchanger, which transferred the heat to fuel coming from the fuel tank at a temperature of 300°F. The fuel temperature would be raised slightly to 315°F and would then be used to cool the 1250°F ram/bleed air down to 325°F. At that temperature it could be used to power the air conditioning system.

Convair began a design study for the air conditioning system, with a subcontract award expected by the end of June. The system would use a water boiler to cool the 325°F air to 110°F. This air would spin dual turbines each driving an air-conditioning compressor. Given the extreme temperature of FISH, a redundant system was essential to protect the pilot and equipment.

Because the turbines would heat the air, it would be cooled again to 110°F in the boiler and fed to the compressors, where it would emerge at −50°F. It would be routed through the sections of the aircraft needing cooling and dumped overboard. The cabin and electronics bay would be kept at 70°F, the reconnaissance payload at 110°F, the wheel wells at 200°F, the auxiliary power unit and air-conditioning equipment bays at 275°F, and a wiring tunnel at 400°F.

Another subcontract was with Minnesota-Honeywell (M-H) for an automatic flight control system. M-H would do an initial design study and work with Convair to prepare a procurement specification. The spec was to be ready by the end of June [110]. The system would have a sophisticated self-calibration feature to eliminate the need to manually change the gains in the feedback system. It would be a fly-by-wire (FBW) system with direct mechanical backup in case the FBW system failed. To reduce that possibility, the electronic channels would be duplicated. The hydraulic controls would also be duplicated. Test systems would be delivered in April and May of 1960, with the first production units at the end of October.

Autonetics, a division of North American Aviation, would prepare the specification for the navigation system. The plan was that it would be an adaptation of the AGM-28 Hound Dog air-launched guided missile's digital inertial navigation system and have a target accuracy of 1 mile per hour of flight. It would use a model N5G inertial platform that was linked with a Verdan MBL-9A computer. At Mach 4, that was an error less than 0.035%. It would operate independently of the B-58's navigation system, requiring ground alignment before takeoff. An optical sight would be included for the pilot to take fixes and observe the area below him. The first production units would be delivered in August 1960.

New FISH

Gene Kiefer met with Convair to discuss a six-week program extension during which they would take the design of FISH to the level of detail needed for production. In addition to the subsystem work that had been contracted out, Convair itself would be doing additional work on improving RCS, lift/drag ratio, structures, and ram recovery, as well as ongoing work with Marquardt on the ramjets and ducts. New models would be needed, and existing ones would be modified to test the new designs (Fig. 78). Work on

Fig. 78 FISH desk model. (Courtesy of Allyson Vought and Chad Slattery. Copyright © 2009 Chad Slattery.)

Fig. 79   Three-quarter view of the FISH wind-tunnel model. (Courtesy of Lockheed Martin.)

materials and manufacturing techniques would continue. And, finally, more detailed project plans and cost estimates would be done [111].

Approval of the program extension was presumably contingent upon being given a go-ahead following the Land Panel review on 9 June.

By the end of May, Convair had finished the redesign of FISH. They had completed almost 300 hours of testing of 1/17th-scale models (see Figs. 79 and 80). There had been 30 hours of testing of the composite (mated) configuration (see Fig. 81) at speeds up to Mach 2 and 57 hours on FISH alone at those speeds. During March, the FISH model had also been run for a total of 200 hours at speeds from Mach 2 to above Mach 4 in a 4-ft unitary wind tunnel. The inlet itself had received 100 hours of tunnel testing at speeds up to Mach 5 at the Jet Propulsion Laboratory (JPL), in Pasadena, California. A 1/10th-scale model of the ramp leading to the inlet was also tested in a Convair wind tunnel.

Convair had assumed responsibility from Marquardt for testing models of the inlet ducts; in exchange, Marquardt had transferred $150,000 to Convair.

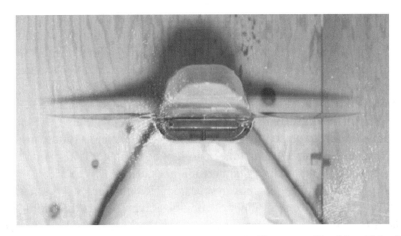

Fig. 80   Front view of the FISH wind-tunnel model. (Courtesy of Lockheed Martin.)

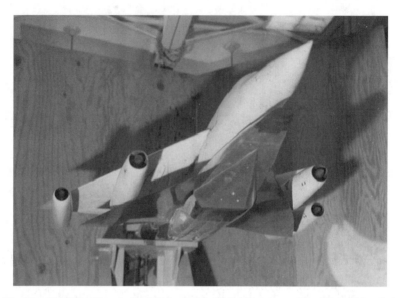

Fig. 81   FISH/B-58 mated wind-tunnel model. (Courtesy of Lockheed Martin.)

By the end of April, about half the money had been spent and $74,000 was returned to Marquardt [112].

The result of aerodynamic and radar testing had proved the inlet in several ways. It had been shown to have a low radar return. The inlet provided adequate pressure recovery at varying angles of attack, and the ducts delivered air to the ramjets with little distortion across the face of the engine. Finally, Widmer's team had obtained preliminary inlet control data.

Testing of the composite configuration revealed three aerodynamic problems (Fig. 82). First, the coefficient of drag was nearly double that of the B-58 alone. This meant that although the B-58 could accelerate from subsonic speed to Mach 2 in about three minutes, with FISH attached it would take almost nine. Second, FISH had to be lengthened to improve stability and balance, as well as to provide space for more fuel. Finally, the center of pressure of the composite was found to be farther forward at low speeds than anticipated. To remain stable, the center of gravity would have to move forward, and that could only be done by lengthening the B-58 by 5 ft.

This meant that the originally planned B-58A carrier, with its J79-5 engines, would have to be replaced by the proposed—but as yet unfunded—B-58B, with its higher-power J79-9 engines and longer fuselage. It also would have beefed-up wings and landing gear and a larger tail.

FISH had also changed in ways besides lengthening its fuselage (see Fig. 83). Externally, one of the most obvious changes was the addition of a canopy to give the pilot direct vision rather than requiring him to rely on television cameras. The twin vertical stabilizers were moved from midway

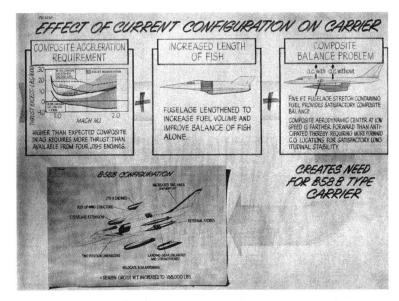

Fig. 82   Lengthening FISH and B-58 carrier. (Courtesy of the Jay Miller Collection and the Aerospace Education Center.)

out on the wings to the sides of the rear part of the fuselage. The rear skids were replaced with wheels, at a cost of 388 lb. Unfortunately, the payload decreased from 560 to 415 lb. A 60-lb optical sight was added. And 201 lb of radar-absorbent material was added to the inlets and 155 lb to the rest of the airframe. The gross weight went from 35,027 to 38,325 lb.

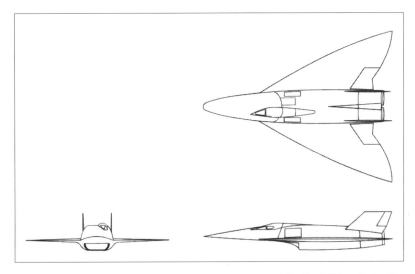

Fig. 83   FISH general arrangement drawing. (Courtesy of the Jay Miller Collection and the Aerospace Engineering Center.)

Wind-tunnel tests had shown that above Mach 2, FISH had a pitch-down moment of 0.01. To compensate would have required nose-up trim, which would have increased drag and reduced range. To change the pitch moment to near zero and reduce the trim needed, the nose and wing tips were cambered up, giving the bottom of the fuselage and wings a slight saucer appearance.

Other, less obvious changes included adding cooling water and more oil for the turbojets, while reducing the amount of oxygen for the pilot. On the negative side, the L/D of the wings had decreased from 5.9 to 5.4. Despite the additional fuel, the increased weight and reduced L/D cut the overall range from 4150 to 3900 n miles, 2.5% below the target of 4000 miles.

Using that range estimate, Convair had constructed various mission profiles. Without refueling, the B-58 could carry FISH about 900 n miles to its launch point. With one or two refuelings, this could be extended to either 2100 or 4200 miles. The missions would take off from either London, England, or Fairbanks, Alaska, and fly about 300 miles outside Soviet or Chinese territory to avoid radar until the launch point was reached. Launches over the Arctic Sea would traverse the Soviet Union and land in Karachi or Okinawa. Launches near Japan would traverse China and fly along the Sino–Soviet border, landing in Karachi. Launches over the Mediterranean would also land in Karachi. FISH would make at most two 90-deg turns during the flight. Virtually the entire Soviet Union and China could be covered with one mission profile or another.

The propulsion system also changed. The single buried JT-12 was replaced by a pair of J85s that would pop sideways out of the fuselage behind the cockpit (see Fig. 84). This solved a number of problems, including the need to modify the JT-12 to tolerate the extreme temperatures that it would experience between the ramjets. The Y-shaped converging duct design was eliminated, as the engines would be out in the airstream to the sides of the upper part of the fuselage. This position gave better inlet and exhaust performance and also improved accessibility for maintenance. Finally, moving the engine forward improved aircraft balance. The main cost was an increase in weight of 272 lb for the nacelles and the engines themselves.

The ramjet nozzles had also changed to improve their efficiency and reduce their RCS. The original nozzle design contained a plug in the form of a cone pointed forward. As the exhaust gases flowed through the nozzle, they were compressed between the plug and the surrounding cylindrical shroud. This gave an efficiency of 0.96, whereas an efficiency of 0.98 would add 290 nautical miles to FISH's range. Finally, while the flat back of the plug gave an acceptable radar return, it required that the shroud be built of dielectric material, which proved troublesome.

The new design substituted a plug that gave the appearance of two cones placed base to base. The first (convergent) part compressed the exhaust gases

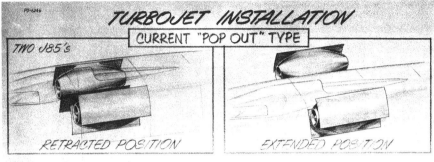

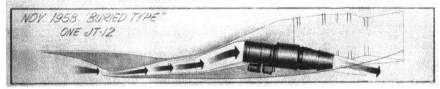

Fig. 84 Change in FISH turbojet installation. (Courtesy of the Jay Miller Collection and the Aerospace Education Center.)

as before, but the second (divergent) part was lined with absorber, as was the shroud. The nozzle achieved the desired efficiency of 0.98. As of June 1959, there were still problems developing an absorber material that could tolerate the extreme heat, but it was felt that this problem could be solved.

Convair had also done extensive testing of materials for the various components of FISH. These included Iconel-X and Rene-41 steel alloys in a variety of configurations, including tensile test coupons, brazed honeycomb panels, and panel splices. Experimentation with other alloys proved the feasibility of tolerating 1250°F temperatures. Pyroceram specimens were produced as 3×15-in. beams in thicknesses from 0.15 to 0.25 in., as well as sheets and wedges. These were proven to meet all of the strength and durability requirements.

The manufacturing team under Vinco Dolson constructed a full-scale 112×46-in. wing box of steel honeycomb to prove the brazing technique. The oven used could accommodate a B-58 elevon and could heat half of FISH's wing in one piece. A 67-in.-long by 54-in.-deep notched leading edge was also produced and subjected to assembly and structural tests.

Today, a small part of that wing material is all that remains of FISH.

Chapter 12

LOCKHEED'S LARGE AIRPLANES

After having been unable to design either a stealthy or nonstealthy small aircraft that could meet the mission requirements, Lockheed returned to the original concept seen in Archangel I, an aircraft designed purely for performance and having no concessions to reduced RCS. The A-10 and A-11 were the two designs that resulted from this new philosophy.

THE A-10

With the A-10 (Fig. 85), Lockheed was able to reduce Archangel I's takeoff gross weight by 16,000 lb, increase the midmission altitude by 2500 ft, and increase the cruise Mach from 3.0 to 3.2. This was based upon using the J93 engine, which was under development for the B-70 bomber. When the design was started, the J93 was thought to be ahead of the J58 in development. However, after further investigation, it turned out that the J93 was actually 18 months behind the J58.

The original A-10 design made no concessions to reduced RCS. It was designed to go high and fast. Nevertheless, some attempts were made to evaluate how difficult it would be to make it stealthy (Fig. 86). The small-scale RCS model had RAM added to slope the sides of the fuselage and the engine nacelles. Although this might have improved the RCS, it was probably only a small amount because the corners between the nacelles and wings and between the fuselage and wings still remained. The effect of the increased drag on performance was also probably negative. However, by moving the engine nacelles away from the fuselage, the corner airflow problem seen in Archangel I was avoided.

With a cruise Mach number of 3.2, an altitude of 90,500 ft, and an operational radius of 2000 n miles, the A-10 could meet the performance objectives for the mission. However, because the J93 engines were lagging in development, a new design using the J58 would be needed.

THE A-11

In March, Johnson's team went to work on what they thought would be their final major design. The operational concept of the A-11 would be to fly

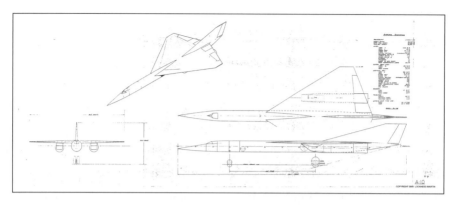

Fig. 85 A-10 general arrangement drawing. (Courtesy of Lockheed Martin.)

out of the Zone of the Interior (ZI), that is, the continental United States. Using two midair refuelings, it could have a total range of 13,340 miles over the course of an eight-hour mission and avoid the security and political problems of operating out of a foreign base.

The original A-11 design had a dual fuel system, carrying 31,000 lb of HEF and 17,000 lb of JP-150. Two J58 engines would burn the HEF at cruise altitude and speed and the JP-150 at lower and slower flight regimes. The rectangular inlets for the J58 engines were designed by Ben Rich. To improve pitch stability, at one point the team considered using a canard.

In mid-March, Ed Baldwin began laying out the A-11 (Fig. 87). With a length of 117 ft and a span of 57 ft, it was larger than the A-10. It was 6130 lb heavier, of which 2630 lb was additional fuel. It could deliver the same 2000-mile cruise radius but at 93,500-ft altitude, 3000 ft higher. Henry Combs

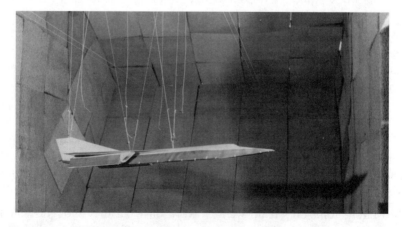

Fig. 86 A-10 model with RAM treatment in original anechoic chamber. (Courtesy of Lockheed Martin.)

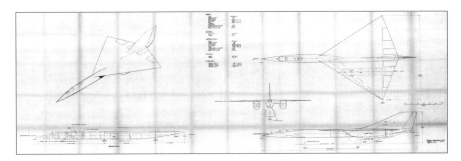

Fig. 87 A-11 general arrangement drawing, by Ed Baldwin. (Courtesy of the family of Edward P. Baldwin.)

and other engineers wrote SP-114, "Proposal: A-11," and published it on 18 March. It was delivered to the Agency (CIA) for analysis. (Private communication from Chris Pocock.)

Baldwin worked through a number of other variations, including the A-11A, which used J93 engines, and in April a set of wing-tip fins, which would have run from the leading to the trailing edge just outboard of the elevons and extending 45 in. above and below the wing (Fig. 88).

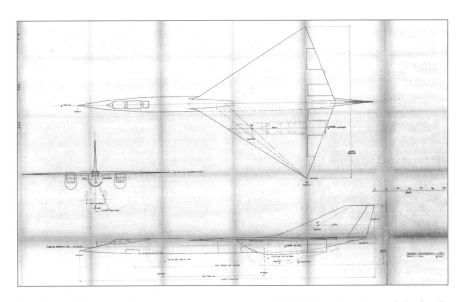

Fig. 88 A-11A general arrangement drawing, by Ed Baldwin. (Courtesy of the family of Edward P. Baldwin.)

Chapter 13

JUNE 1959 LAND PANEL REVIEW

By the end of May, the operations staff had completed their staff study comparing the A-11 and FISH. On 5 June, William Burke published a memo, GUS-0285, which compared the aircraft. Its conclusions were as follows:

1) Both aircraft could cover all the targets in the Soviet Union, Soviet satellites, and China.

2) Both were equal in range, altitude, and sonic boom. FISH was "a little better" in radar detection, but more vulnerable to infrared detection. With its speed 400 knots faster, FISH was less vulnerable to intercept.

3) Because the A-11 could operate from the Zone of the Interior, it would have better operational security. Because the B-58 launch aircraft was a nuclear delivery system, FISH presented "… an almost insurmountable security problem."

4) The A-11 was thought to be much easier and economical to maintain.

5) The cost of the A-11 would be much less than that of FISH.

6) FISH would require eight times as many trained people to operate as would the A-11.

Burke recommended that the A-11 be chosen [113].

On 9 June, Johnson and Widmer and members of their teams once again went to Boston to brief the Land Committee. In preparation, Lockheed had prepared SP-120, "Operational Analysis: A-11," as well as their own stealth study, SP-119, "Probability of Radar Detection of Airborne Targets," which was also known as the "Cat and Mouse Study" (Private communication from Chris Pocock).

After the meeting, Johnson wrote that,

> I gave the A-11 pitch and reported on about six months of radar studies which we had made, in which we proved, at least to ourselves, that improvements available to radars at the present time would enable detection of any conceivable airplane which would fly in the next three to five years. We specifically computed that the probability of detection of the A-11 was practically 100%.
>
> I think I made some kind of an impression with the radar people, because the ground rules changed shortly after this and it was agreed that the A-11 would make such a strong target that it might be taken for a bomber [114].

Johnson returned to Burbank thinking that Lockheed was out of the competition.

Widmer presented the new FISH design and compared it with the one from November 1958. He reported that if given a go-ahead at the end of June 1959, major assembly would begin in February 1960, and the first flight would be at the beginning of 1961, an 18-month schedule. By the middle of the year, training of operational pilots could begin, by which time three aircraft would be on hand. The 12th and last FISH would be delivered in November 1961 (see Fig. 89).

The schedule assumed that the B-58B program would be approved almost immediately, that J79 engines would become available by the end of 1959, and that the first test flight would happen in July 1960. The number 68 and 79 aircraft would then be made available in January and March 1961 as the two carrier aircraft for the program.

To date, the development costs were just over $2 million. Over the next 30 months, $64 million would be spent on the first three FISH articles and two B-58B carriers, and between March 1960 and the end of 1961 nine more articles would be delivered at a cost of $33.5 million. Neither of the estimates included the ramjets from Marquardt. Convair would receive a fee of 6%.

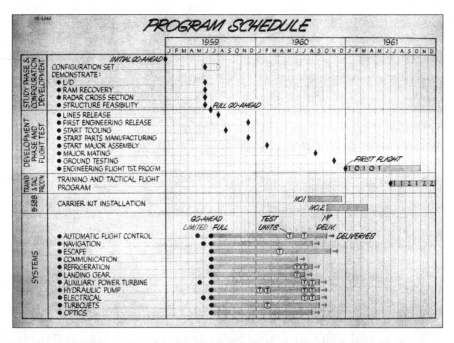

Fig. 89   FISH production schedule. (Courtesy of the Jay Miller Collection and the Aerospace Education Center.)

According to Bissell's remarks at the 16 June staff meeting,

> The Boston meeting did not result in a decision regarding the two vehicles. It was agreed that from an operational viewpoint the A-11 with its 4100 nautical mile range, and increased altitude capability, was highly desirable. Other features are its ability to use a short field for take-off and a common fuel. It also has the advantage of unassisted take off. Its major draw back is that it will be constantly tracked. The Convair vehicle, being a staged type, has this built-in draw back; on the other hand, there is more possibility of flying missions undetected than with the A-11 as presently designed. Basically, the technical experts at Boston on the panel were concerned only with aircraft perfection and radar cross section and not operational use. The meeting came to one conclusion and that was the sporadic detection and tracking by radar must be expected regardless of vehicle [115].

DEMISE OF *FISH*

The final nail in FISH's coffin came near the time of the Land Panel review in June 1959, when Widmer went with a Convair delegation to present the B-58B to General Curtis LeMay, the vice chief of staff of the Air Force (Fig. 90). He was accompanied by J. T. McNarney, the president of Convair; August Esenwein, Widmer's boss; and R. C. Seybold, vice president of

Fig. 90   B-58B/B-58MI desk model. (Courtesy of Roger Cripliver.)

engineering. McNarney wanted to sell the B-58B, but he also wanted to recover the investment in the B-58A, which would require selling three more wings of B-58As before they could afford to begin shipping the B-58B.

The Air Force audience included LeMay and a general whom Widmer remembers as "Shorty" (probably Maj. Gen. Hewitt T. "Shorty" Wheeless, director of plans in the office of the Deputy Chief of Staff for Plans and Programs). Widmer had known both men for some time. LeMay kept a special sports car at Offutt Air Force Base, powered by a turbine engine provided by Boeing, and Shorty had once let Widmer, himself a sports car enthusiast, drive it.

Widmer stood alone at the front of the auditorium and gave the presentation. The B-58B had features that LeMay wanted, such as side-by-side seating for the pilot and copilot. From his World War II experience flying bombers over Europe, he felt that this gave superior crew coordination compared with tandem seating, where the crew members could not see what each other were doing and had to rely only on the intercom and on instruments that might not agree.

LeMay was pleased with what he heard and asked, "When can I have this?" McNarney fielded the question, telling LeMay that the B-58B would be available after SAC had taken three more wings of B-58As.

LeMay said nothing, but stood up and walked out of the auditorium. Everyone was shocked into silence, but Shorty knew his boss and knew what his leaving meant. He told Widmer, "I want you to come to my office." Alone there with Widmer, he said, "I'll tell you: your company will never do business with the Strategic Air Command" (Interviews with Robert H. Widmer, Fort Worth, TX, 13 March 2003 and 7 Nov. 2003).

The B-58B was dead. And without a launch vehicle, so was FISH. In November 2003, Widmer's parting words on FISH were "I wonder if it would have worked."

CHANGING DIRECTIONS

On 17 June, Bissell was scheduled to meet with Air Force Generals Holzapple and Demler to agree on a consistent position between the CIA and the Air Force on which vehicle to recommend to the Director of Central Intelligence and to the Chief of Staff of the Air Force. Either the meeting did not take place, or they did not reach a final decision because they would meet again on 8 July. Nevertheless, a consensus was emerging that both Convair and Lockheed would have to redesign their aircraft again.

Lockheed was instructed to redesign the A-11 into a configuration that would incorporate RCS reduction techniques, even at the expense of cruise altitude. It was a vindication of sorts of their conclusion of six months earlier that stealth and maximum performance were mutually incompatible. It was a reasonable tradeoff, though, in that the reduced RCS might make up for the slightly longer time the aircraft would spend in the acquisition radar's target zone. For Convair, however, the change was much larger—a repudiation of the complexity of the parasite concept despite FISH's much lower RCS. Convair was directed to redesign FISH as a single-stage aircraft using a pair of J58 engines.

On 23 June, Kiefer reported at the weekly staff meeting that Convair was continuing to pursue FISH while starting work on their single-stage design. He also said that Lockheed had agreed to study reducing the RCS of the A-11. Neither company had agreed to report in by a certain date, although Kiefer added that 15 July was the effective deadline because certain equipment would not be available after that date [116].

On 1 July, the program office sent formal instructions to Convair:

> Contractor will study the feasibility of a new vehicle configuration utilizing only turbojet engines, having a gross weight approximately 100,000 pounds, and capable of take-off without a carrier aircraft.
>
> A. Analysis and Design—An analysis and design program will be carried out on the above described configuration type to study the feasibility of this configuration to meet the requirements for electronic effects and performance as specified in document [redacted].
>
> B. Models and Components—Small scale electronic models will be constructed and tested to study the Rodger's [sic] Effect on the new configuration.
>
> C. Materials Research—A study will be conducted of materials research requirements for the above described configuration.
>
> D. Subsystems—Subsystem requirements for the new configuration will be analyzed [117].

On 3 July, Bissell visited Lockheed to let Johnson know that he was still in the competition. Johnson later wrote, "just at about the time when I thought we were ruled out, they extended our program and agreed to take lower cruising altitudes which we could obtain with a version of the A-11 adapted in shape and treatment to reduce the cross-section" [118]. Work began immediately on turning the A-11 into the A-12.

When Bissell and the Air Force representatives met on 8 July for a final decision, "It was reported that Kelly Johnson has agreed to revise his design to reduce radar return. Convair now has configuration on the test at Indian Springs. Marquardt has completed the first engine for tests. Mr. Bissell is expected to report to the President the middle of this month" [119]. They decided that "the Kelly Johnson proposal sounded best and would be supported in the meeting with Dr. Killian" [120]. Word got back to Johnson, who wrote, "As of July 8th, it seems that there is a good chance that, if an airplane will be built for the mission, it will be ours" [118].

KINGFISH

Convair suddenly had an enormous amount of work to do in only a few weeks. Although the change from A-11 to A-12 would be a significant amount of work for Lockheed, it would nevertheless be only an incremental change in a design that had been evolving for a year. Convair, however, had to design an entirely new aircraft from scratch. The new design was called KINGFISH.

Widmer's team quickly worked through a number of different configurations for the new aircraft (Fig. 91). Initially, the new design looked like a larger version of FISH, but with a wide, flat-sided fuselage containing the J58 engines. The bottom of the wings and fuselage was a smooth slightly convex surface, and the leading and trailing edges of the wings were curved—an adaptation of the flying saucer. Rectangular inlets were located on the sides of the fuselage behind the cockpit and ahead of the leading edges; the ejector was a pair of single expansion ramp nozzles (SERNs) with a low vertical partition between them. A bump in the nozzle helped prevent radar energy from entering the ejector. The canopy was still offset toward the port side of the fuselage, as it had been on the June FISH design. The twin vertical stabilizers extended directly up from the sides of the fuselage, on either side of the ejectors, and sloped slightly inwards. Although this arrangement would have reduced the amount of infrared (heat) energy that was emitted to the sides and it would have helped keep radar energy out of the ejector, it also would have required careful design to prevent weakening of the stabilizers by the heat of the exhaust.

Fig. 91 Early KINGFISH desk model resembled FISH, but with rectangular inlets and bifurcated SERN exhaust. (Courtesy of Allyson Vought and Chad Slattery. Copyright © 2009 Chad Slattery.)

Fig. 92 This KINGFISH design has a wider fuselage and more-prominent inlet bleed. (Courtesy of Allyson Vought and Chad Slattery. Copyright © 2009 Chad Slattery.)

A minor variant of that design widened the fuselage and removed the partition between the ejectors (Fig. 92).

A third variant used a V tail mounted in between the ejectors (Fig. 93). This was probably rejected because the heat of the exhaust would have weakened any structure that could have been built at a reasonable weight.

Fig. 93 KINGFISH with V-tail. (Courtesy of Allyson Vought and Chad Slattery. Copyright © 2009 Chad Slattery.)

Fig. 94 KINGFISH with vertical stabilizers on wings. (Courtesy of Allyson Vought and Chad Slattery. Copyright © 2009 Chad Slattery.)

A fourth design had more extensive changes (Fig. 94). The fuselage was widened, the sides were sloped inwards, and the canopy was moved to the centerline of the fuselage. The partition between the ejectors was deleted, and the vertical tails moved $\frac{3}{4}$ of the way to the wing tips, so that the leading edge of the tail met the leading edge of the wing. This design would probably have had greater rudder authority in case of engine-out operation.

The fifth variant added a third vertical tail, an extension of the partition between the ejectors (Fig. 95). The other two were moved $\frac{2}{3}$ of the way out from the sides of the fuselage toward the wing tips. The leading edge of the vertical tails began about 2 ft behind the leading edge of the wing.

The final design was similar to the fifth variant, except that the center vertical tail was deleted, leaving the low partition between the ejectors (Fig. 96). The partition probably improved single-engine performance by keeping the exhaust directed to the rear. Without the partition, the exhaust from a single engine would have spread toward the centerline, increasing the tendency to yaw into the dead engine. Also, the bases of the two remaining vertical stabilizers did not reach the trailing edge of the wing, and their trailing edges were swept forward. By keeping the stabilizers away from the edges of the wing, the wing provided better shielding from radar. The bottom of the aircraft was convex, approximating the "flying saucer" shape.

Convair needed a full-scale RCS model, and they needed it quickly. Rather than build one from scratch, they decided to modify the FISH RCS model. This had the disadvantage of being smaller that KINGFISH would be, and so the model would not actually be full scale, but "large scale." This meant that

Fig. 95 KINGFISH with three vertical stabilizers. (Courtesy of Allyson Vought and Chad Slattery. Copyright © 2009 Chad Slattery.)

radar frequencies would have to be adjusted upward and that the response of the radar-absorbent material in the design would not exactly match that in the final design. The new inlets would also require careful study.

Convair laid out the plan in a 16 July message to Kiefer:

> In attempting to follow the plan of marking time on FISH and proceeding with KINGFISH we find that the latter is difficult to do effectively without

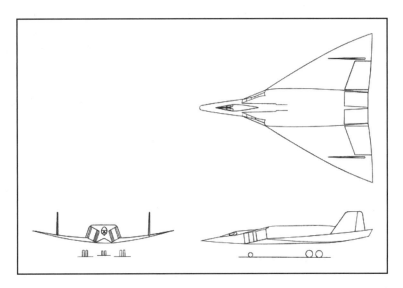

Fig. 96 Final KINGFISH general arrangement drawing. (Courtesy of the Jay Miller Collection and the Aerospace Engineering Center.)

some degree of electronic testing. We have therefore, been searching for a program which is compatible with our current funding situation and which would give us valuable information on the KINGFISH program [121].

The plan was first to study the advantages of KINGFISH's smooth bottom. Rather than building a full-scale model from scratch, the inlet section would be removed from the full-scale FISH radar model and the opening faired over. This 7/10-model would be studied at S band at a variety of angles before and after modification.

The second step was to study $\frac{1}{8}$th-scale inlet models. The "Herring" inlet cowl would be modified to simulate Fiberglass construction, and an egg-crate grill would be simulated in the subsonic area of the inlet. It was hoped that this would solve a problem that had been seen at 70 MHz. To simulate 400-MHz radars, measurements would also be made at S band. If time permitted, similar modifications and tests would be run on the "Smelt" inlet. The total effort was small—four people for one week [121].

The next day, the chief of the DPD contracts branch sent a telex advising Convair that "contract with [Marquardt] for ramjet was terminated effective 17:30 hours 15 July 1959. [Convair] should govern its contracts and design efforts accordingly" [122]. Convair replied that "all aspects of the low-speed wind tunnel tests currently in work at Langley Research Center which relate specifically to FISH alone have been stopped. Only those portions of the tests which are applicable to KINGFISH are being continued" [123].

On 29 July, Convair sent a message to Gene Kiefer proposing two plans for the work they would do between 27 July and 30 August. Plan 1 would only include one week's RCS testing of the large-scale KINGFISH model. Because it did not include any wind-tunnel testing, if Convair were selected, then the desired start date of 9 November would slip by three to five weeks. Essentially, most of Convair's team would be doing almost nothing for most of August. Plan 2, on the other hand, was aimed at doing as much work as possible in preparation for a start date of 9 November. It included a wind-tunnel model of the pre-inlet, material purchasing, structural testing, and subsystem studies and layout [124]. On 31 July, Kiefer spoke with Convair and approved Plan 1 [125]. Formal approval would not come until mid-August under Technical Change Request number 13 [126].

The FISH model was taken off the pole at Indian Springs and shipped to Fort Worth. The model had a metal frame covered first with a thin translucent covering and then with metal foil (Fig. 97). The skin was stripped, and the framework changed to the new shape. The ventral inlet was removed, and the fuselage was reshaped to include the rectangular inlets for the J58s. The vertical stabilizers were removed from the sides of the fuselage, and new ones were placed on the wings, held in place with wires (Fig. 98). As with FISH, the wing edges would be serrated to receive RAM inserts (Fig. 99).

Fig. 97 The full-scale FISH RCS model being re-worked into the large-scale KINGFISH model, before addition of reflective foil. (Courtesy of the Jay Miller Collection and the Aerospace Engineering Center.)

Fig. 98 KINGFISH model showing vertical stabilizers. (Courtesy of the Jay Miller Collection and the Aerospace Engineering Center.)

Fig. 99 Triangular teeth were assembled separately before being attached to the wing edges. (Courtesy of the Jay Miller Collection and the Aerospace Engineering Center.)

At the same time the RCS model was being modified, Convair also built a wing-like shield for the pole at the range to reduce reflections from it (Fig. 100). The top of the shield was attached to the top of the pole with a swivel; at the bottom was a set of removable wheels. As the pole with the model was raised, the shield would be pulled upwards until the pole nestled inside the shield. Then a hinged section on each side of the opening would close around the back of the pole.

In barely over two weeks, the modified model was back on the pole at the Indian Springs Air Force Base antenna range for measurement. It was mounted belly up on the pole and apparently measured without the wedges of radar-absorbent material in the leading and trailing edges. The absence of the wedges would give a worst-case measurement that could be corrected mathematically.

The Indian Springs facility also featured a small railroad track that ran parallel to the direction of the radar beam and about 20 ft to one side. The track carried a corner reflector on a pole and would move the reflector at one meter per second. Reflections from the model that hit the corner reflector would be reflected back at the model and then back to the radar receiver. The additional data allowed the engineers to triangulate on the exact locations on the model that were sources of large reflections. This was more accurate than if they had only measured the simple reflection from the model to the receiver. (See Figs. 101 and 102.)

Fig. 100 Construction of shield for pole to support large-scale RCS model. (Courtesy of the Jay Miller Collection and the Aerospace Engineering Center.)

Fig. 101 KINGFISH model on pole at Indian Springs, with "railroad" visible on left. (Courtesy of Lockheed Martin.)

Fig. 102 Model with shield attached to pole and ready to be lifted. (Courtesy of Lockheed Martin.)

On 5 August, Edgerton, Germeshausen, & Grier (EG&G) requested authorization to spend $2000 to build an inflatable bag to support the Convair model. It would be closed with a helical seam; because the seam would spiral around the bag, it would have a smaller reflection than one running vertically. The money would be provided under EG&G's existing contract number TE-2191 [127].

To get the testing done in time, Convair wanted to conduct measurements seven days a week beginning 15 August and running through the end of the month. EG&G passed the request for overtime to the Agency [128]. EG&G proceeded with measurements using the 70-MHz radar. The results turned out to be similar to those that Lockheed would achieve with its modified design.

THE A-12

Dick Fuller was tasked with laying out the A-12 (Fig. 103). L. D. MacDonald and Ed Lovick provided advice on how to modify the A-11 design to reduce its RCS. There were several stealth features, some more obvious than others. The chines from the nose to the wings were an example of Frank Rodgers's iron maiden shape, which had first been tried on the A-4. The chines merged into the wings, and—to avoid corners—the wings were blended into the fuselage and nacelles. The underside of the forebody was a convex shape and somewhat resembled the early flying saucer. The twin tails were tipped in at 15 deg; this allowed the airplane to bank up to 15 deg before the tails would

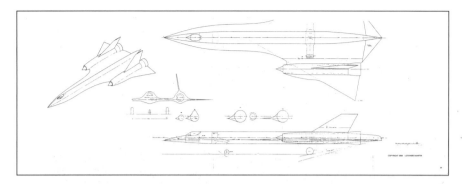

Fig. 103 General arrangement drawing of original A-12 design. (Courtesy of Lockheed Martin.)

reflect a radar signal directly back at a station to its side. (Within the chines, the fuselage was circular in cross section; see Fig. 104.)

Less obvious was the use of radar-absorbent materials. The chines and the leading and trailing edges of the wings were notched and filled with graded dielectric material, Ed Lovick's invention which had appeared in GUSTO 2 and 2A, FISH, and KINGFISH.

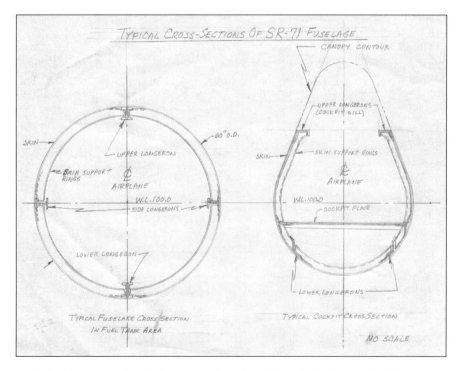

Fig. 104 Fuselage and cockpit cross sections (less chines) for SR-71 and A-12. (Courtesy of the family of Edward P. Baldwin.)

The rectangular inlets of the A-11 were replaced with round (axisymmetric) ones, with a movable "spike" to control airflow. Whereas the spike helped to mask the compressor face, the inlet nevertheless incorporated a great deal of RAM to reduce the energy that did make it all of the way to the compressor. The tails—which at first were fixed with movable rudders and later changed to all-moving rudders—were made of composite materials.

The planform of the airplane was designed to keep the shock wave from the nose from impinging on the rest of the structure of the airplane (Fig. 105). If the shock wave were to make contact, then drag and heating of that area would be increased. If the speed were increased to about Mach 4, the shock would reach the engine inlet and disturb the airflow.

Some of the stealth features also had aerodynamic advantages. Supersonic aircraft experience a change in the location of the center of lift as they accelerate to and decelerate from supersonic flight. As speeds increase, the shock wave off the wing moves farther aft, and the changing pressure distribution behind the shock moves the center of lift. If the center of lift moves too far behind the center of gravity and the elevators lack enough authority to compensate, then the nose will drop, and the aircraft becomes uncontrollable. Even if the elevators can compensate, flight becomes very inefficient. Careful weight management during flight—such as by moving fuel among tanks—can help to keep the center of gravity in the right place without requiring large displacements of control surfaces.

The A-12's chines acted as a fixed canard and developed lift. This meant that the center of lift at low speeds was farther forward and the degree to which the total center of lift moved as the aircraft accelerated was reduced. This reduced the amount of trim drag and gave a L/D of 6.5 at Mach 3.

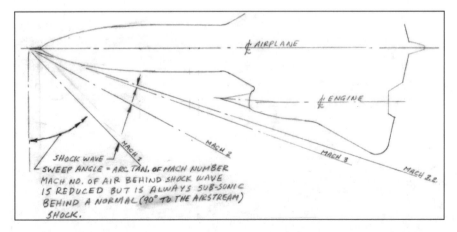

Fig. 105 Shock wave positions at various Mach numbers, by Ed Baldwin. (Courtesy of the family of Edward P. Baldwin.)

The chines also affected the yaw, or directional, stability. If the nose of an aircraft moves to the left or right of the direction of flight, the airflow will exert a force tending to push it farther to the side and requiring correction by the rudders. In the case of the A-12 at a positive angle of attack—and the aircraft always flew at a positive angle of attack at Mach 3—the force was greater without the chines than with them, meaning that the chines improved the stability. A side effect was that smaller rudders could be used, which meant less drag.

The tilting in of the vertical tails helped align them with the airflow off the forebody and leading edges and thus reduced drag. The tilt also caused the side force as a result of deflection of the rudders to pass almost directly through the center of gravity of the aircraft. Had the tails been vertical, the force would have been directly horizontal, several feet above the center of gravity; this would have caused a much larger rolling moment, requiring deflection of the elevons to correct, and increasing drag. Of course, when Ben Rich explained this in an article in 1974, he mentioned none of the RCS effects [129].

In some places the A-12's shape actually violated some of the rules laid down early in Project GUSTO by the SEI team. Viewed from the bottom, the blending of the wings into the nacelles and body formed concave areas. However, these areas would only have been seen as concave if a radar were looking directly up, which radar dishes do not normally do.

The leading and trailing edges of the vertical stabilizers were straight lines, which apparently was not deemed to be a problem, probably because they were made of composites. The initial wing design had straight edges on the leading and trailing edges outboard of the nacelles, which met in a point at the wing tip. The tips were later changed to be rounded.

MINORITY REPORT

By the second week of July, the project office had preliminary RCS estimates for both the A-12 and KINGFISH. Looking at the numbers, Gene Kiefer felt that the redesign was a waste of time because they would probably be detected. He felt that there were two sources of risk in making either the A-12 or KINGFISH hard to track. One was uncertainty about the capability of the Soviet BARLOCK radar; the other was the ability of either Lockheed or Convair to get their RCS below 1 m^2.

There were two competing analyses of BARLOCK. One by the Air Technical Intelligence Center (ATIC), at Wright Field, judged it to be less effective, whereas one by OSI and SEI judged it to be more effective. By ATIC's estimate, FISH would be almost undetectable, and the A-12 or KINGFISH would have a 50/50 chance of detection. By the OSI/SEI analysis, FISH would have closer to a 50/50 chance of detection, and the A-12 and

KINGFISH would be almost as easy to detect as the A-11 or the B-47. Kiefer also pointed out that as long as the B-70 and F-108 were ongoing projects, the Soviets would try to develop countermeasures against these aircraft, which would fly at Mach 3 and 65,000 to 75,000 ft. Those countermeasures would be more effective against the A-11/A-12/KINGFISH than against FISH, flying at Mach 4 and 90,000 ft.

Nevertheless, Kiefer felt that the A-11 was the best bet and wrote that "I fear that the A-12/KINGFISH direction merely delays A-11 by three or four months. The price of useable radar cross section, I fear, is operational complexity. If this price is too high, we should get on with A-11 post-haste" [130].

Resurrecting FISH

Despite the redirection, Convair tried to keep FISH alive. In a 13 July memo, Gene Kiefer wrote that "Convair people do not believe that the decision to cancel B-58B is firm. Their story is that B-58B costs were questioned by Gen. LeMay. Gen. Mark Bradley is to be in Ft. Worth next Friday, 17 July to examine Convair cost estimates" [131]. Convair proposed building a small number of B-58Bs, either by modifying a programmed but unbuilt B-58A or by modifying an existing B-58A. The former would have cost $2.075 million, of which $1.6 million was for higher-thrust J79-9 engines; the latter would have been more expensive as it would have required partially disassembling existing aircraft, for a cost of $6.02 million. They also suggested a third alternative, a six-engine B-58A using less-powerful J79-5 engines, for an incremental cost of $2.63 million.

Kiefer also reported on modifying the J58 engine to work at a top speed of Mach 3.5, rather than 3.2. He reported that the three aircraft would have been able to cruise 3000 ft higher. It would mean a six-month delay with the first flight of the selected aircraft—whether it was the A-11, the A-12, or KINGFISH—slipping from January to July of 1961 [131]. This proposal for Mach 3.5 operation appears to have been based mainly on discussion with Pratt and Whitney, with little or no input from Lockheed or Convair. It is unlikely that all three aircraft would have had exactly the same increase in altitude had their designers worked through the numbers. Interestingly, years later a Lockheed study of pushing the Blackbird to Mach 3.5 revealed a larger number of changes that would have been required, including modifications to the inlet spikes.

More Agency Studies

On 14 July, Bissell and the Land Committee met with James Killian. The various designs were summarized in a memo from Bissell.

Designation	A-11	A-12	KINGFISH	FISH
Mfgr.	LAC	LAC	Convair	Convair
Speed	M 3.2	M 3.2	M 3.2	M 4
Alt. start cruise	86,500 ft	83,000	84,000	90,000
Range, n miles	4,100	3,940	4,070	3,900
Gross weight	94,500 lbs	110,000	101,000	38,300
Length, ft	106	100	78	47
Wing span, ft	56	56	52	37
Powerplant	2-J58A/B	2-J58A/B	2-J58A/B	2-Ramjets
2-J85				
1st flight	Jan 1961	Jan 61 (metal) May 61 (radar matls.)	May 61	Jan 61

The A-11 design by Lockheed represents an attempt to obtain the highest level of aerodynamic performance without recourse to operational complexity beyond conventional aerial refueling and is uncompromised by unusual features intended to minimize detection by radar. This design is backed by about two months of study and low speed wind tunnel tests only.

The A-12 and KINGFISH designs are of most recent vintage. These designs attempt to retain operating simplicity in addition to incorporating features to minimize their radar echo. Only a few radar model tests at 70 mc. have been accomplished on the KINGFISH version in the short time since these designs were started. Hence the estimates of aerodynamic performance and other characteristics have not as yet been substantiated by tests and detailed study. There is very little difference between these two designs at the present time.

The FISH proposal represents the design approach recommended in the 15 November report to you. This design is a modification of the original SUPER HUSTLER concept. The aircraft is carried aloft and accelerated to supersonic speed by a B-58 mother craft.

Wind tunnel model tests have demonstrated the validity of the estimated aerodynamic characteristics. However, the more powerful engines scheduled for the B-58B series aircraft are needed for acceleration. Structural testing has established confidence in the materials of construction. Radar testing including a full scale model has established that the aircraft should have an exceedingly small radar echo at frequencies near 70 megacycles, 600 megacycles, and S-band. While it would be desirable to further reduce the radar cross section the amounts and extents of the higher-than-desired radar echos [sic] are relatively small [132].

The memo then summarized the optimistic radar analysis by ATIC and the more pessimistic one by the CIA. It went on to say that S-band testing had not been done on either the A-12 or KINGFISH, and that while their RCS could not be reduced to the level of FISH's, "... it could be made sufficiently low so as to make radar tracking extremely difficult" [133]. The memo stated that if the Soviets were to develop an airborne infrared detection system, then any of the four aircraft could be detected. It also estimated that although there were theories of sonic-boom propagation, there were little practical data; the designs were estimated to generate booms at irregular intervals along perhaps half of their ground tracks.

> All four of these aircraft are estimated to have sufficient range to reach all Sino-Soviet territory. The A-11, A-12, and KINGFISH designs could operate from a single base in the U.S. with rendezvous refuellings outbound and inbound from KC-135 tankers based in Alaska and Greenland. A third rendezvous refueling would be needed with a tanker based in North Africa for the deepest penetrations. The FISH aircraft would require bases in Alaska and either Greenland or England and buddy refueling of the B-58 mother from a KC-135 on longest missions. Landing facilities for the FISH and for a cargo aircraft for retrieval would be needed at three locations near Soviet territory or the FISH could be towed to home base by a KC-135 as in a prolonged rendezvous aerial refueling.
>
> In recent days the continuance by the Air Force of the B-58B airplane and that of the J79-9 engine has become quite unlikely. Since the B-58B aircraft is needed to accelerate the FISH to supersonic speed in order to launch, the cancellation of the B-58B together with the operational complexity of the FISH proposal lead to the conclusion that further consideration of the FISH is unwarranted. Similarly, due to the conventionally high radar echo expected from the A-11 design further consideration of this proposal is unwarranted.
>
> Approximately three to four months of testing and study may be needed to establish the same level of confidence in the estimates of radar echo, aerodynamic performance, and other characteristics of the A-11 or KINGFISH designs as is now held in the case of the FISH proposal. It is recommended that approval be given to undertake the necessary tests and detailed study at an estimated cost of $1,750,000. Because of the similarity of designs this additional work would be undertaken with but one and not both of the contractors now in this program [132].

FRAMING THE DECISION

On the 18th, probably to get their ducks in a row for a discussion with the President, Bissell's staff completed a summary of the GUSTO alternatives. They considered detailed range estimates for different mission profiles. In particular they considered the effect on the ultimate penetration range of the aircraft, depending upon how many times it was refueled and whether the refueling aircraft (a KC-135) would return to its home base or land closer to the denied territory, an option with political implications. Refitting KC-135s with J57 turbofan engines to extend their range was considered as a way to avoid their having to land at a foreign base.

The range analysis was also done for FISH and its B-58 carrier. If launched by a B-58B that was refueled once, the launch point would be 3900 miles from base, with the B-58B flying 2000 miles back to another base for landing. If a specially built six-engine B-58A with its higher fuel consumption in the mated configuration were instead used as a launch aircraft, the launch point would only be 2500 miles out in order for the B-58A to have the same 2000-mile return range. If a more distant launch point were desired, more refuelings would be needed. Refueling of FISH itself was even considered. Because of its small fuel tank—the J85 engines were really only intended to flatten the landing approach, rather than dead-sticking after ramjet shutdown—its range after refueling would be only 250 miles. However, if FISH were to stay on the KC-135's boom and continuously take fuel, it could be "towed" up to 3000 miles. The memo did not mention possible exhaustion of the FISH pilot during the six hours of the extreme concentration needed to fly while on the boom.

The memo provided cost figures for 12 aircraft for the various options. The A-11 would be $82.5 million, and the A-12 with its antiradar treatment would be $84 million. KINGFISH would be $152 million. J58 engines for any option would be $72.5 million, giving totals of $155 million, $170 million, and $224.5 million. FISH would be $97.5 million for the 12 airframes and equipment and $47 million for its ramjets and turbojets. Three six-engine B-58As would be only $6.5 million, whereas three B-58Bs—assuming that the Air Force canceled it and the Agency would have to pay for development of the airframes ($36 million) and the J79-9 engines (another $36 million)—would be either $151 million or $216.5 million.

There were several recommendations. To achieve the earliest possible operational readiness, Lockheed should be given an unlimited go-ahead, which would shortly see them spending $1 million per month. If no major changes were required based on RCS analysis, then an all-metal aircraft could fly in January 1961 and the number two aircraft—with RAM—could fly in May. Operational readiness would be achieved in the spring of 1962.

It was recommended that Convair continue design, wind-tunnel, and radar work on a "more cautious" basis, spending $1.75 million over four months. If the results at that point looked good, then they would be given a go-ahead (presumably if Lockheed were experiencing difficulties) and could have a first flight in May 1961, with operational readiness in the spring of 1962. One option to keep the program on track would be to abandon the antiradar requirements.

The J58 program was more problematic. As funded by the Air Force and Navy, it was currently on a "... starvation diet and scheduled to close down

December 1959" [133]. Bissell's staff felt that if funding were increased to $2 million per month, Pratt and Whitney could produce a Mach 3.2 bypass engine by January 1961 and a Mach 3.5 engine by mid-1962.

The memo also contained an enigmatic reference to a possible FISH variant:

> Fan engine program is a possibility. Pratt & Whitney expect to run a fan engine version in November (CONFIDENTIAL info.) which would contribute to M 4 design to fit enlarged but unstaged FISH configuration. Most optimistic guess is 2-1/2 to 3 years to first flight. Probably program costs $175 million airframe plus $100 million engine development if built by Convair/P&W. This becomes interesting only if low radar return is still mandatory and cannot be achieved in larger aircraft [133].

It is unclear to which engine this referred. The intention seems to be a version of FISH that could take off under its own power and still cruise at Mach 4.

Perhaps to anticipate a question instigated by the Air Force, the memo evaluated using either the B-70 or the F-108 as a reconnaissance platform. Because the B-70 was so large and unstealthy, it was dismissed with "one must answer affirmatively to this question in order to consider the B-70 any further: 'If there were no U-2 aircraft today, would the B-52 be considered for the U-2 mission?' " [133]. The answer was obviously a no.

The F-108 had actually been considered by the Agency. It was about 25% heavier than the A-11 and would have to be stripped and redesigned to meet the range requirements, but would still have an altitude 10,000 to 15,000 ft lower. It would not be at all stealthy. Competition with the Air Force for airframes and J93 engines would mean that operational readiness would be at least four-and-one-half to five years away, to say nothing of the security issues of working within a normal Air Force program [133].

INFORMING THE PRESIDENT

On 20 July, Allen Dulles led a delegation to the White House to brief the president on the choice for the follow-on vehicle. The group included General Cabell, Bissell, Secretary of Defense Neil McElroy, General White, George Kistiakowsky, and Killian. The story recorded by Goodpaster was a condensed form of Bissell's memo to Killian, but with some changes. The A-11 and KINGFISH were not mentioned at all. The A-12 was revealed to have an expected RCS 1/20th that of a B-47. The A-12 was described as having a speed of Mach 3.2 initially and 3.5 in a later version. As far as detection of the A-12, "We must anticipate that the presence of the aircraft would usually be known, although there would be a great deal of confusion arising from its height and speed, and there is very little likelihood that successful tracking could be carried out" [134].

They pointed out the security and political advantages of operating it only from North American bases.

There was a discussion of times and costs. Either FISH or the A-12 could fly in January 1961. The A-12 would cost $170 million and FISH $160 million (including converting B-58s). This would require $100 million in FY 1960; $75 million had been earmarked and was available from Department of Defense (DoD) funds and $25 million was available from the CIA.

Allen Dulles commented that the A-12 could be modified for use as a bomber, dropping bombs from 90,000 ft; General White questioned the practicality of the idea.

Eisenhower approved the program, saying that the development should be kept moving, both for peacetime and wartime reconnaissance. He noted that with $6.5 million spent to date on the Convair proposal and $100,000 on Lockheed's, "the outcome was an expensive lesson that the latter is the more promising." He said that "we want to have the finest reconnaissance aircraft that we can provide" [134]. He directed the group to investigate funding and to have close technical supervision. That apparently raised the specter of too much oversight because there was a discussion after which he agreed to stick to the pattern of development of the U-2 [134].

FINAL DECISION

On 20 August, both Lockheed and Convair submitted their final proposals. Lockheed presented the A-12, and Convair presented KINGFISH. Compared with the estimated performance numbers presented in July, the range of the A-12 had improved, and that of KINGFISH had decreased. Overall, the A-12 had a significantly better range and a better altitude over most of the cruise phase. In his 1968 history of the OXCART program, Johnson wrote that the RCS of the two aircraft was the same. He also stated that KINGFISH used "plastic"

FINAL COMPARISON OF A-12 AND KINGFISH [CIA]

	A-12	Kingfish
Speed	Mach 3.2	Mach 3.2
Range (total), n miles	4,120	3,400
Range (at altitude), n miles	3,800	3,400
Cruising altitude		
Start, ft	84,500	85,000
Midrange, ft	91,000	88,000
End, ft	97,600	94,000
Cost of 12 aircraft, excluding engines	$96.6 million	$121.6 million

afterburners that "... could not conceivably work" [74]. He is also reported to have criticized KINGFISH's inlets as being very sensitive to unstarts.

It was 28 August before Lockheed got the news that they had been selected. Johnson wrote in his log that he,

> Saw the Director of the Program Office alone. He told me that we had the project and that Convair is out of the picture. They accept our conditions (1) of the basic arrangement of the A-12 and (2) that our method of doing business will be identical to that of the U-2. He agreed very firmly to this latter condition and said that unless it was done this way he wanted nothing to do with the project either. The conditions that he gave me were these:
>
> 1. We must exercise the greatest possible ingenuity and honest effort in the field of radar.
>
> 2. The degree of security on this project is, if possible, tighter than on the U-2.
>
> 3. We should make no large material commitments, large meaning in terms of millions of dollars.
>
> We talked throughout the day on problems on security, location, manpower, and aircraft factors. At noon I took 9 of the project people out for luncheon, in celebration of our new project [135].

Despite the celebration, the approval had come with one difficult condition. Lockheed had until the end of the year to prove that they could reduce the A-12's RCS (Fig. 106).

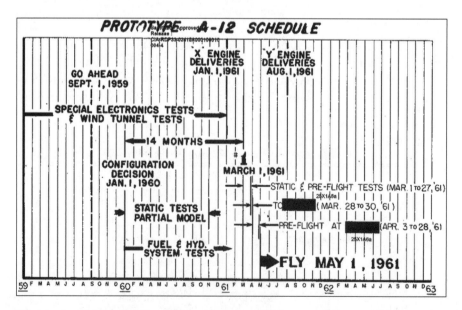

Fig. 106 A-12 production schedule. (Courtesy of the Central Intelligence Agency via the National Archives and Records Administration.)

Ramping Down Convair

As a hedge against problems with the A-12, the Agency continued to fund Convair's work at a low level. At first this included a three-month wind-tunnel study of the "two-dimensional" single expansion ramp nozzle from mid-September to mid-December 1959. Convair would also prepare a summary report on KINGFISH and submit it on 11 October. Next, they would prepare a report on the application of the "Rodger's Effect" to both the FISH and KINGFISH designs. The report was to include the results of additional tests on the large-scale and $\frac{1}{8}$th-scale KINGFISH models. Finally, if the tests on the two-dimensional nozzle were successful, then Convair would design—but not test—a wind-tunnel model of the inlet [136]. This work was approved by the Agency. The two-dimensional exhaust would be tested first in Langley Research Center's wind tunnel and then in the Jet Propulsion Lab's supersonic wind tunnel. The final report would be ready by 31 January 1960 [137].

On 11 September, Convair advised the project office that they were proceeding with construction of the exhaust nozzle using corporate funds and requested that the Agency arrange for testing at JPL to begin on 9 November [138].

With the reduction in scope of the project, Convair had access to radar-absorbent materials and steel honeycomb that were now unneeded. These were either on hand in Fort Worth or in transit from suppliers. Following discussion with Gene Kiefer, at the end of September Widmer proposed that the materials be used for a research program during October and November. For the RAM, they would investigate the effects of high temperatures and of temperature cycling and would also try to find the best ways to attach the RAM to the wings. The cost of the studies would come from money saved by canceling material orders and subcontracts. He also proposed to dispose of excess steel and Iconel-X by either selling the raw materials and returning the money to the Agency or by shipping the materials to the Agency [139]. This proposal was also approved, with completion due on 18 December 1959 [140].

Convair's work came to an end in February 1960, when Bissell's deputy chief of the development branch telephoned Convair and "informed him that all GUSTO feasibility study work being performed on our behalf by Convair is hereby terminated." Arrangements were made for either Kiefer or Parangosky to visit Convair and discuss the termination and any open contract matters [141]. Following the meeting, various reports were made, including a final patent report with the interesting statement that "no 'Subject Inventions,' reasonably appearing to be patentable, were conceived or first actually reduced to practice in the performance of the aforementioned contract" [142].

The chief of the DPD contracts branch visited Marquardt on 29 March and Convair in Fort Worth on 30 March to discuss the details of closing out the project. Marquardt had raw materials, parts and components, tooling, and two completed ramjets, all of which had to be disposed of. Items that were sensitive from a security point of view would be scrapped, and items that were not would be transferred to other contracts if possible. A final report was produced, and a "sterile" version of that was sent to Col. Norman Appold, at Wright Field [143].

In addition to parts and tooling, Convair also had wind-tunnel and electronic models in Fort Worth, as well as models and other materials at Indian Springs. Bissell was concerned that the KINGFISH large-scale model at Indian Springs, as well as all of the radar measurement data on it would be retained "for possible use in the unlikely contingency that the A-12 develops serious electronics troubles and that we wish to demonstrate to Lockheed the electronic performance achieved with an alternate configuration" [144].

In Fort Worth, radar models and sensitive parts and tooling were to be destroyed, although the wind-tunnel models could be retained. They were allowed to retain technical data, for future work, but in a sanitized form. Some office equipment, like calculators, were shipped to Lockheed [145]. Convair also delivered its "KINGFISH Summary Report," and it was judged by the Technical Analysis Staff to be "satisfactory evidence of completion of the basic contract plus changes" [146].

By the beginning of June, Convair was well underway with the sanitization process and requested a visit by the Agency to approve what had been done [147]. The ground rules for the inspection were laid out by the chief of the technical analysis staff. There was to be no mention that the CIA had sponsored the work, no mention of the RCS reduction work, and no mention of potential reconnaissance missions. Convair pushed back on destroying the RCS data because they felt that it would be valuable in future contracts and losing all of the lessons learned would put them at a disadvantage. It was decided that it could be stored in a secure facility and not used, pending a policy decision [148].

The inspection took place on 27–28 June. The engineering data to be retained were "quite voluminous" and still contained the code names FISH and KINGFISH. Specific data about the FISH and Super Hustler air-launch techniques were to be destroyed, as was information on KINGFISH missions, probably because the performance of KINGFISH was close enough to that of the A-12 so that it would have revealed its capabilities and missions. Convair agreed that before it would submit any future proposals that used the antiradar data or the performance data, that it would ask the Agency whether it desired to review the proposal [149].

A year and a half later, the president of Convair Fort Worth formally released the government from any further liability. A caveat allowed Convair to bill for any third-party liabilities (such as subcontractors or suppliers) that were unknown at the time [150]. Convair also signed over any refunds, rebates, and credits to the government [151]. It would be another year before all of the finances had been wrapped up and Convair had presented all of the necessary documents to the Agency [152]. By then, Richard Bissell had resigned, and his Development Projects Division had become part of the newly formed Office for Special Activities (OSA). Finally, on 18 January 1963, the final payment was made to Convair [153]. Project FISH was finally over, almost five years after Bissell's first visit with Widmer.

Ten months later, the FISH design came out of the files, to be used as the starting point for a proposal for a follow-on to the A-12. Widmer assigned Randy Kent to run the new project, designated only by the internal billing reference, Work Order 540. Work started the week of President Kennedy's assassination in November 1963. The design was similar to FISH, but with variable-sweep wings. Like FISH, it used Marquardt ramjets burning hydrocarbon fuels. It would have had a range of 4000 n miles. Work continued until the following July, but with no funding in sight, the project was cancelled (Interviews with Robert H. Widmer, Fort Worth, TX, 13 March 2003 and 7 Nov. 2003). The FISH data ended up in two locked file cabinets that Widmer last saw when he left Fort Worth for a two-year assignment in S. Louis. When he returned, the file cabinets were not to be found. Presumably, the Agency had collected them.

OXCART

Even before the final decision was made, the Agency chose a name for the upcoming project to build the follow-on aircraft. On 31 July, the project security officer explained in a memo that the name GUSTO had become widely used in both the government and industry. There was no need for anyone who would not be involved in the building and operation of the follow-on aircraft to know that GUSTO had been anything more than a feasibility study. Those who would not be briefed into the new program would be told nothing at first; GUSTO would be phased out over three to four months and eventually terminated for budgetary reasons, giving the impression that the study had come to nothing [154].

The name for the new project would be OXCART.

A month later, William Burke sent Leo Geary a draft of a statement that would explain the cancellation of GUSTO to any Air Force personnel who were aware of it. He requested that a senior commander would issue the statement that "the development of a Super-Hustler reconnaissance aircraft" would be "indefinitely deferred." The reason was to be budgetary reductions, which resulted in the cancellation of the B-58B. It concluded stating that "Each individual must ... consider the program temporarily cancelled rather than abandoned and must continue to maintain its highly classified status" [155].

Geary rejected the idea of formal debriefings of Air Force personnel. He told the project security officer that they "would only create more problems than currently exist. The number of Air Force personnel knowledgeable of the Project are relatively few and approximately one-half of the GUSTO cleared Air Force personnel are scheduled to be briefed on the follow-on OXCART Project" [156].

In December, John Parangosky was appointed coordinator of OXCART, which meant that he had the responsibility to ensure that the program advanced efficiently without duplication of effort. He had worked for Bissell for some time on the U-2 and would eventually be honored by the Agency as one of the pioneers of overhead reconnaissance. William Burke explained to the DPD branch chiefs that in this new position Parangosky had been given the authority to call on them for assistance as needed [157].

Chapter 14

PROVING THE A-12

Although Lockheed had won the competition, they now had to prove that the A-12 could actually be made stealthy. Measurement of a full-scale RCS model would give the proof needed, but there was a great deal of invention and analysis to be done. On 31 August, Lockheed began work on the full-scale model, a $\frac{1}{8}$-scale model, and a pole (or elevation post) on which to mount the full-scale model (Fig. 107).

The site for the testing had to be selected. On 10 October Johnson, along with Leo Geary, visited Indian Springs, where the U-2, FISH, and KINGFISH had been tested. In his first progress report on Project OXCART, Johnson reported that

> we were aiming at a completion date of 12 October when we discovered that there was no hope of putting the model on the [Indian Springs] pole. Even if there were, it was very apparent when we visited the place that we would be taking extreme liberties with over-all security if we put the big thing at the [Springs] at all.

So he changed the completion to the week of 16 November and planned on hauling it to the Ranch, where it would be mounted on a new pole.

They then proceeded to the Ranch and met Frank Rodgers to view the proposed site for the measurement range. Johnson was

> totally unimpressed with the area that had been proposed for installing the new range ... There were numerous ditches, ten-foot piles of gravel, depressions, and other impediments, which did not improve conditions over those at the [Springs]. In talking to the group, I proposed that we swing the whole range out onto the west end of the lake, so that the pole would be a few hundred yards from the west shore, but with an absolutely level shot from a new building on the southwest corner that could be obtained and provide ideal conditions.

The pole (or "elevation post") was built from three destroyer propeller shafts, welded end to end. It was designed by Lockheed's Henry Combs, who many years later would design the titanium claw used by the CIA's Glomar

Fig. 107 Full-scale RCS model of A-12, showing wood framework. (Courtesy of Lockheed Martin.)

Explorer to lift a Soviet submarine from the floor of the Pacific Ocean (Interview with Henry G. Combs, Santa Clarita, CA, 19 April 2003).

Installation of the pole for the model was the critical item needed in order to be ready for arrival of the model in mid-November. The pole and model would be lifted into the air by a large piston. To give the concrete in the base time to cure before installing the piston, the concrete would have to be poured by 2 November. To speed things up, John Parangosky verbally approved Lockheed's designing and building the piston and the rotating head that would cap the pole and hold the model. The one-month delay would give Lockheed time to put the foam RAM in the chines on one side of the model; by comparing measurements of the two sides, the effectiveness of the RAM could be determined.

While waiting on the work at the Ranch and on the full-scale model, other work could go forward. By the end of September, a $\frac{1}{8}$-scale model would be tested. A 10-ft section of a full-scale fuselage would be used to test various chine arrangements. Johnson expected that by the end of October they would have tested a full-scale nacelle with various inlet configurations.

From measurements of the U-2, Johnson knew that the engine ejectors would produce a large reflection. He had once proposed to Bissell that they make RCS measurements of an F-104 with its engine running, both with and without afterburner. Now with the coming availability of the new pole, he suggested mounting an F-94C, "which has one of the world's biggest, noisiest afterburners, and which can readily be changed in its fuel mixture setting to vary from no external flame to one about 25-ft long" and a thrust up to 9000 lb. This would allow them to evaluate whether there actually would be a cone of ionization that would mask the ejector.

Lockheed's focus was almost entirely on the RCS techniques and measurements. To supplement the staff in Burbank, Johnson proposed to borrow four to six engineers from another division for up to a year. He was also in consultation with the subcontractor who would design and build the vertical stabilizers and would provide them with the tooling to produce them. Other design work was moving forward on the fuel system, landing gear, and wing and fuselage. With good progress being made, he informed Bissell that he would be going to Hawai'i to "study the aerodynamics of surfboards and the big canoe with outriggers as it comes in past Diamond Head" [158].

At one point in the experiments, Johnson asked Lovick whether it was necessary to have the chines on both sides when measuring the 10-ft fuselage section. Lovick answered "no," which turned out to be incorrect. With no chines on the side away from the radar unit, an effect called "creeping waves" appeared and caused increased reflections. When the chines were applied to the far side, the effect disappeared, and the measurements were more realistic (Interview with Ed Lovick 8 March 2008).

The plans for the new range at the Ranch were ambitious. EG&G was responsible for building the facilities, running the radars, performing maintenance, and—at an office in Las Vegas—performing data reduction. Under the ongoing contract TE-2191, they requested permission to build a 50-ft pole shield ($8,000), a 165-ft railroad and tractor with a 50-ft retractable mast ($5,000), a 55-ft retractable mast for calibration ($2,800), and dual recording equipment that would allow making simultaneous measurements at two frequencies, thus speeding up the tests [159].

Meanwhile, Lockheed's L. D. MacDonald set forth his own requirements for changes to be made to the equipment being moved from Indian Springs to the Ranch. He was trying to get more consistent measurements that could be easily compared with results from Lockheed's anechoic chamber. He suggested changing the amplifier and recorder system to eliminate nonlinearities, which would eliminate doing corrective hand computations and would also make plots more easily compared. They would also standardize on a polar log chart for displaying data, install equipment for 70-MHz measurements, and use timers to warm up equipment before it would be needed. Although the 70-MHz measurements would potentially have accuracy problems, they were the only way to accurately measure ferrite materials in full-scale models because no way had been found to scale the materials down for measurement in a small model. Finally, MacDonald reiterated EG&G's requests to simultaneously test at both S and L bands and to improve the pole arrangements, noting that the existing 36-in. sphere was severely damaged and needed to be replaced with a new 42-in. one [160]. The work ramped up quickly, and by

1 October Lockheed had requested that EG&G go on a six-day-week schedule to support the preparations for testing [161]. SEI's Ed Rawson designed much of the radar instrumentation and supervised its installation. Lockheed's Leon Gavette designed the rotator for the pole.

On 2 November, Johnson and Bissell met to discuss project status and management. The objective was to review the schedule and the status of the many activities that were in progress. Initially, both teams met together to review the overall program. This included how Lockheed's and Pratt & Whitney's schedules would mesh, why the A-12 with radar-absorbent materials would fly later than the all-metal aircraft, what had to be done before the configuration freeze date of 1 January, what reports were to be rendered, what should be covered in each report, and what was the possibility of adding a synthetic aperture radar and infrared sensors in addition to the camera.

The meeting then broke up into technical and administrative sessions. The technical session covered test results on RCS, wind-tunnel models, engine and nacelle design and RCS of the ejector, speed improvement from Mach 3.2 to 3.5, temperatures expected, and airframe loads from maneuvering and from encountering gusts and wind shear. It also covered materials (RAM, plastics, and titanium), the camera environment and window design, the pilot environment (including using a pressure suit and an ejection seat vs a capsule), fuel status, and various subsystems (inlet controls, air conditioning, guidance, and communications).

The administrative session covered security matters such as cover stories to vendors and clearances for personnel, competitive bidding, effects of the steel manufacturing strike, transportation of personnel and articles to the Ranch, runway requirements, improvements to facilities, and general logistics [162].

Despite EG&G's efforts at the Ranch, when Johnson followed up the meeting with a letter summarizing the discussions, he noted that, "I am very perturbed that the completion of the new site will be delayed at least two weeks beyond our original plans. In any case, we will be ready with our test models" [163]. He included an outline of a test plan that he'd reviewed with Frank Rodgers and reiterated the need for additional equipment to keep up the speed of the work and to get better data [163].

The original test plan consisted of 17 test programs, numbered 0 through 16, each with its own objective (see Table 1). In each program, measurements would be made at UHF, L band, and S band. The A-12 models to be tested included a $\frac{1}{8}$-scale model, a full-scale 10-ft-long fuselage section, and the complete full-scale model. The chine and wing edge treatments tested would include teeth with 110Ω foam with "tapered paper" (TP), teeth with 110Ω foam with varying numbers of layers of Teledeltos paper (TD), teeth with 2-in. "subteeth" on the edges of the teeth, resistive honeycomb material, and ferrite beads to supplement the foam. The $\frac{1}{8}$-scale model and the 10-ft-fuselage model would also be mounted vertically for some tests and rotated 360 deg to

TABLE 1 Test Plan

Number	Title	Objective	Model configuration(s)
0	Site illumination test	This series of tests, to be performed by [redacted] and LAC jointly, is intended to verify the suitability of the target illumination or to aid in correcting defects that might exist	The area in which the target will be illuminated shall be scanned with the aid of antennas suitable to the frequency employed. An adjustable antenna tower mounted on a low-silhouette vehicle shall be used to support the exploring antenna
1	Hydraulic ram (pole) tests	To determine the return from the hydraulic ram (pole) and its protective treatment	50-ft ram with its protective treatment
2	Bag and jig tests	To determine the return from the bag and support jig for the $\frac{1}{8}$-scale model	The 22-ft bag with spiral seam with and without the jigs that support the $\frac{1}{8}$-scale model: a) horizontally (6 deg) and b) vertically (nose up)
3	Standard pattern series (full scale)	To provide a set of standard reference patterns from an all-metal version of the A.C. at all frequencies and angles available at the [redacted—Ranch]	Full-size aircraft aluminum covered chines, aluminum inlet diffuser, aluminum-covered fins tilted 15 deg inboard, aluminum-covered wings. aircraft to be tested upside down on the pole
4	Standard reference series ($\frac{1}{8}$-scale)	To provide a set of standard reference patterns from an all-metal version of a $\frac{1}{8}$-scale model at all frequencies and angles available at the [redacted—Ranch]	One-eighth scale aluminum-covered chines, aluminum-covered inlet diffusers, aluminum-covered fins tilted 15 deg inboard. (Check of [redacted] work.)

(*Continued*)

TABLE 1 TEST PLAN (CONT)

Number	Title	Objective	Model configuration(s)
5	Basic treatment tests (full scale)	To determine the effects of the basic (low temperature) anti-radar treatment. Basic treatment means that treatment best known as of 2 Nov. 1959	Full-scale aircraft.: a) right-hand side all metal; left-hand side treated with 110Ω foam + T.P. in chines and wings. 110Ω foam in left-hand inlet diffuser all-metal fins, tilted 15 deg inboard b) same as a) except for fins tilted 10 deg inboard c) same as a) but with plastic fins tilted 15 deg inboard d) same as a) except that fins shall be removed e) same as a) except for plastic fins at 0 deg
6	Basic treatment tests ($\frac{1}{8}$-scale)	To determine the effects of the basic (low temperature) anti-radar treatment using 110 med. polyether foam and T.P. and to provide attenuation data for frequencies equivalent to some that are not available at the [redacted—Ranch]	$\frac{1}{8}$-scale aircraft.: a) right-hand side all metal; left-hand side treated with 110Ω foam + T.P. in chines and wings 110Ω foam in left-hand inlet diffuser, all-metal fins tilted 15 deg inboard
7	Protection angle-basic treatment ($\frac{1}{8}$-scale)	To investigate the broadside return as a function of angle above and below the horizontal plane. To determine the amount, if any, and the angle of protection provided by the chines	$\frac{1}{8}$-scale: a) right-hand side all metal; left-hand side treated with 110Ω foam + T.P. in chines and wings 110Ω foam in inlet diffuser, all-metal fins tilted 15 deg inboard
8	Standard reference series (10-ft fuselage section)	To provide a set of standard reference patterns for all frequencies and angles available at the [redacted—Ranch]	10-ft fuselage section: a) right-hand (by definition) side all metal; left-hand side treated with 110Ω foam + T.P. in chine section mounted with its axis horizontal b) same as a) except the axis shall be vertical and the section shall be rotated about its axis

9	Afterburner Outlet configuration ($\frac{1}{8}$-scale)	To determine how much, if any, difference in return results from the use of different afterburner outlet configurations	$\frac{1}{8}$-scale: a) circular outlets, cut off $\perp$ to centerline of nacelle b) circular outlets, cut off 80 deg to centerline of nacelle c) outlets of b) rotated in 90 deg steps d) square outlets, cut off $\perp$ to centerline nacelle e) square outlets cut off 80 deg to nacelle (tilted 10 deg forward) f) square outlets of e) rotated in 90 deg steps g) rectangular outlets of 1.5:1 anti-radar cut off $\perp$ centerline of nacelle h) rectangular outlet of g) cut off 80 deg to centerline nacelle i) outlet of h) rotated in 90 deg steps
10	Nacelle tests ($\frac{1}{8}$-scale)	To determine the degree to which the nacelles contribute to the total response on a treated model	$\frac{1}{8}$-scale model: a) all-aluminum, complete model with the nacelles removed b) completely treated model: 110Ω foam + T.P. in chines and wings, aluminum fins tilted 15 deg inboard, 110 deg foam in inlet diffusers; 110 deg foam + T.P. in nacelle chines c) same as b) except nacelles removed and gaps smoothed with aluminum foil
11	Subteeth tests (10-ft section)	To determine the effect, if any, of the addition of subteeth to the edges of typical fuselage chine teeth	10-ft fuselage section: a) right-hand side aluminum covered; left-hand side treated with 110Ω foam + T.P; 2-in. subteeth added to each edge of the chine teeth on the lower surface of the typical chine b) same as a) except 2-in. subteeth added to top surface if those on the lower surface cause any improvement
12	Doubly tapered T.D. tests (10-ft section)	To determine the effect, if any, of doubly tapering the T.D. stacks on each of the chine gap surfaces	10-ft fuselage section: a) right-hand side aluminum covered; left-hand side treated with 110Ω foam + T.D. stack cut as described in Project Memo No. 8, Sec. 8e

(Continued)

TABLE I TEST PLAN (CONT)

Number	Title	Objective	Model configuration(s)
13	Comparison T.D. vs T.P. (10-ft section)	To obtain information about optimizing the resistance taper required if foam is used	10-ft fuselage section: a) right-hand side aluminum covered; left-hand side treated with 110Ω foam + T.P. stack b) same as a) except using T.P. in place of T.D.
14	Resistance loaded honeycomb	To determine the parameters of the resistance-loaded honeycomb in an effort to replace foam and T.P. or foam and T.D. with a better structural material	10-ft fuselage section: a) right-hand side aluminum covered; left-hand side treated with resistance-loaded honeycomb in the chines gaps b) same as a) but with T.D. added one layer at a time, increasing the loading until a significant change occurs c) same as a) but with 110Ω foam inside the chine gaps to increase the loss
15	Ferrite edge loading	To determine the value, if any, or the penalty, if any, (at high frequencies) of the use of ferrite loading of chine and wing edges	10-ft fuselage section a) right-hand side aluminum covered; left-hand side treated with 110Ω foam + T.D. ferrite beads on a thin (0.040 in.) wire across each chine gap
16	Afterburner evaluation	To determine the effect of hot gases (and ionization) on the return from the tailpipe of a typical jet anticraft	F94-C aircraft on 50-ft pole: a) Engine idling or shut off b) Engine running + afterburner c) Engine running + afterburner + water

A.C. = aircraft; A.R. = anti-radar; C/L = centerline.

simulate rolling the airplane. Finally, ionization of the exhaust would be measured by putting an actual F94-C fighter on the pole and running its engines with and without afterburners and water injection. (This final phase of testing the F-94C was eventually dropped.)

The test of the 2-in. subteeth turned out to be ill conceived. For the teeth to have had a useful effect, their size would have had to be related to the wavelength of a frequency of interest. Because the 2-in. size was an arbitrary choice and was not based on a particular frequency, they had no significant effect and were not retained in the final design (Interview with Ed Lovick, 8 March 2008).

A 22-ft-tall inflatable bag was used to support the $\frac{1}{8}$-scale model. The bag was almost completely transparent to radar and was thus better than a metal pole with a shield. To minimize the reflections from its seams, the seams spiraled around the bag, rather than running straight up its sides. First, the bag would be inflated and a calibration sphere placed on top of it. Because the sphere had a known RCS, the reflections of the bag and surrounding terrain could be measured (Fig. 108). Then the sphere would be replaced with the small model and the measurements would proceed (Fig. 109).

Fig. 108 Inflated support bag with calibration sphere. (Courtesy of the Central Intelligence Agency via the National Archives and Records Administration.)

Fig. 109 A $\frac{1}{8}$-scale model of the original A-12 design was tested on the new bag. Note the spiral seams on the bag that gave a lower reflection than the straight seams on the older bag. (Courtesy of Roadrunners Internationale.)

The spiral seam bag was the third design used for the Lockheed models. The first one, designed by L. D. MacDonald, was a cylinder with a round top and made of translucent plastic. The phallic shape became the butt of numerous jokes, and it was eventually scrapped. The next one was tapered—a cone with a round top—but with straight seams that turned out to cause measurable reflections. After the spiral seam bag went into use, the one with straight seams became known as "the old bag."

Radar-Absorbent Materials

The first full-scale RCS model tested at the Ranch used a foam radar-absorbent material (RAM) for the serrated "teeth" in the chines and in the leading and trailing edges of the wings. The foam was in the form of inch-thick sheets that had been soaked in a graphite suspension, resulting in a constant density of graphite through the material and thus a constant resistivity. The foam was then run through a pair of rollers to squeeze out the excess. (Known as "110Ω foam," its impedance was not actually 110Ω; it was just a name.) The foam came to be called "hockey puck" and the machine the "puckey squeezer." As the end of the foam went through the rollers, they would come together and splatter the suspension onto anyone standing too close. Baldwin remembered that the first time Johnson watched the machine in operation he was with Leroy English, one of the designers, and Johnson had not been warned. Then "the machine spit on his white shirt. He said, 'Leroy, that machine just speckled my brand new white shirt, first time I ever wore it.' Of course the collar was frayed, and all" (Baldwin, E. P., quoted in Baldwin, R. E., unpublished transcript of tour of SR-71 by Ed Baldwin and Lou Wilson, Beale Air Force Base, California, 30 Aug. 1989).

"Tapered paper" (TP) was an asbestos sheet that had been sprayed with a graphite paint to give a conductivity that increased logarithmically from the outer edge of the chine or wing to the inmost edge of the sheet. SEI supplied an alternative, Teledeltos paper (TD). Both TP and TD had the good property of being easy to cut to the desired shape. Each was tested as the surface layer for the teeth in the chines, with foam underneath. Eventually Teledeltos paper was discarded because its resistivity varied too much (Ed Lovick, e-mail to author, 1 March 2008).

The foam itself was just an initial idea and would not work in the actual aircraft, flying at Mach 3.2 and generating tremendous heat that would melt the foam. However, it was good enough for initial testing of an aircraft on a pole or flying at relatively low speeds. The hope was that a Fiberglass-based honeycomb material loaded with graphite would have the same electrical properties as the foam, while tolerating the high temperatures and being strong enough to handle the dynamic pressure from the airstream.

Ed Baldwin remembered that,

> L. D. MacDonald and our chemist Mel George did most of the work on the development of all that material and how to handle it. The use of loaded honeycomb was our idea, and we developed it because we didn't know any other way to get a loaded core that was both lightweight and stiff. We wanted both and honeycomb was a good way to do that. (Baldwin, E. P., quoted in Baldwin, R. E., unpublished transcript of tour of SR-71 by Ed Baldwin and Lou Wilson, Beale Air Force Base, California, 30 Aug. 1989.)

In the final form, the honeycomb was made of asbestos, not Fiberglass. The outer layer of the material was an asbestos matt up to 3/32nd-in. thick, which had been formed and then hardened by baking. Next was a layer of loaded honeycomb made of asbestos and silicone sheeting about 6/1000th-in. thick and then bent into the honeycomb shape that made up the core. To strengthen the edges of the sheets where they were attached to the aircraft, a 1/16th-in.-thick layer was wrapped around the edges. The outer layers were glued to the honeycomb core; the gluing was critical because if any significant area were not properly glued, that area would lack strength, and the surface would fail. The honeycomb core was loaded with tiny graphite balls, about 5/10,000th-in. in diameter. The graphite was mixed with a solution that would make it adhere to the honeycomb.

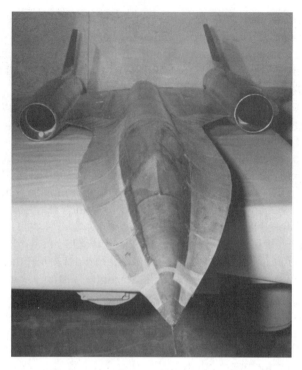

Fig. 110 $\frac{1}{8}$-scale A-12 RCS model with simulated RAM sheets. (Courtesy of Lockheed Martin.)

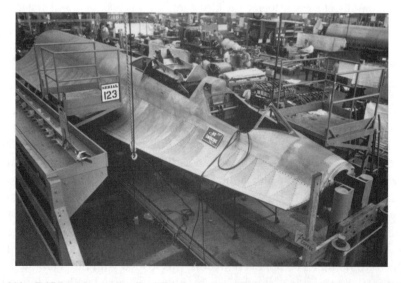

Fig. 111 RAM teeth are clearly visible in Article 123 during manufacturing. (Courtesy of Lockheed Martin.)

Fig. 112 These fragments of resistive honeycomb were recovered from the crash site of A-12 Article 125. The honeycomb piece is $\frac{1}{2}$ in. thick. (Author's collection. Copyright © 2008 My T. Pham.)

Because the honeycomb sheets took time to develop and test, the earliest A-12s (at least articles 122 and 123) were initially manufactured with triangular teeth made of the foam RAM in their chines (Figs. 110 and 111). Each tooth filled the space between the upper and lower surface of the chine. (The prototype, article 121, was never intended for RCS testing and had no RAM in its chines; instead, it used sheet metal triangles for the surfaces.) The chines of articles 122 and 123 were eventually upgraded to 8-ft-long rectangular sheets of the loaded honeycomb having a graded density; this opened up a considerable volume to run plumbing and, in the case of the SR-71, cameras. All subsequent A-12s and all SR-71s used the honeycomb in their chines (see Fig. 112). The wings of all of the A-12s and SR-71s used the foam teeth (Fig. 113).

The manufacturing process also went through a change. At first the density of graphite was varied by soaking different parts of the sheet different lengths

Fig. 113 Article 123's chines would eventually be converted from RAM "teeth" to honeycomb "blankets." (Courtesy of Lockheed Martin.)

Fig. 114   Assembly of A-12 full-scale RCS model. (Courtesy of the Central Intelligence Agency via the National Archives and Records Administration.)

of time. The longer the honeycomb soaked in the resulting graphite suspension, the more graphite would adhere to it. To get a resistivity that varied across the honeycomb sheet, the sheet was dipped in the solution at different depths and for different times; one edge would be lightly loaded and the other edge heavily loaded. There were 11 bands with different densities across the length of the sheet. Later, Lockheed developed a spray-on application for the graphite solution, which gave better control over the grading of the dielectric constant than did the dipping process.

The full-scale model with foam in the chines and wings was completed on 9 November and then hauled in pieces to the Ranch (Fig. 114). The model had a wooden frame with some steel reinforcement; the outer skin was metal foil. It was re-assembled at the site of the pole, which had been lowered into the ground, except for the rotator. On 18 November, the complete model was then mounted on the rotator. A rudimentary shield that had two flat surfaces and that was wide at the bottom and tapered to a narrow top, which hid the rotator, had been assembled. The shield was attached to the pole, and as the model was raised, the shield swiveled up to hide the pole and rotator from the radar set (Fig. 115).

At the end of November, Rodgers sent Kiefer the RCS numbers for the A-12 (Figs. 116 and 117). He summarized the situation saying,

> These represent the recent base situation from which we are proceeding to evaluate addition of vertical stabilizers, inlet spikes, etc. The vertical stabilizers are making a significant increase over these patterns at broadside, but the inlet treatment shows no detectable increase. The 12° curves I

Fig. 115　Mating the full-scale model to the pole and rotator, with shield framework in background. (Courtesy of the Central Intelligence Agency via the National Archives and Records Administration.)

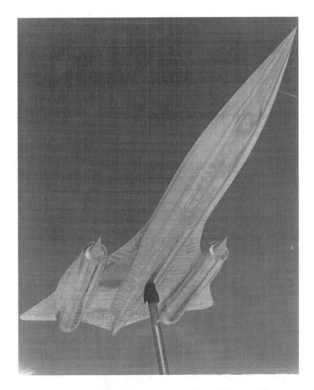

Fig. 116　Full-scale model on pole, without vertical stabilizers. (Courtesy of the Central Intelligence Agency via the National Archives and Records Administration.)

Fig. 117   Original design of fixed vertical stabilizer with rudder. (Courtesy of the Central Intelligence Agency via the National Archives and Records Administration.)

consider the most significant as that is the angle of the potent beam against which we work [164].

The 12-deg curves probably refer to plots of reflection strength when the radar beam struck the model at an angle 12 deg below its horizontal axis. This number comes from two factors. The first is that because of the curvature of the Earth, the radar beam was 4 deg above horizontal at the distance at which it acquired the A-12 as it came over the horizon; if the aircraft were parallel to the Earth's surface, the beam would thus intersect it at 4 deg below its horizontal axis. However, because the A-12 at cruise altitude and speed was expected to have a positive pitch angle (called α or alpha) of 6 to 8 deg, the beam would intersect it at a total angle of up to 12 deg.

One of the early discoveries was that some of the radar beam would reflect down to the ground, back up to the model, and then back to the radar receiver. This negated the effect of reflecting the beam at an "innocent angle." The solution was to cover the ground beneath the model with mats of Hairflex, a horsehair material that absorbed electromagnetic energy. (A similar material had been used by Westinghouse in their early U-2 studies.) To avoid disturbing the mats for minor changes to the model, engineers would be lifted off the ground and over to the model in a "cherry-picker" personnel crane.

It also turned out that the flat-sided shield did not work well. Because the back edges could not be terminated, the electrical currents induced by the radar beam bounced off the edges and caused energy to reradiate back toward the radar unit, interfering with the measurements. Eventually a new shield was designed.

IONIZED EXHAUST

In addition to the passive stealth features of shape and materials, there were some active features. From the early model studies, Lockheed knew that when viewed from the rear by radar, the ejector formed a sort of corner reflector, like a can with one end cut off. Energy entering the ejector would be reflected back at the radar. This was the subject of a meeting at Lockheed with Mel George, Ed Lovick, a physicist working for the Agency named Walters, and others. Lovick suggested that by ionizing the exhaust gases coming from the engine, most of the radar energy could be kept out of the ejectors. Walters agreed, and Lovick had a new project.

Lovick was dispatched to Pratt and Whitney's Florida facility for tests on running engines. He worked there with Harley Nethkin, using a J57 in one of the regular engine test cells. They tried potassium, sodium, and other metallic salts as additives to the fuel. SEI's Dan Schwarzkopf and Ed Rawson used a radar set measuring sideways into the exhaust stream at frequencies of 3 and 10 GHz. To determine how close the transmitter horn could be to the exhaust stream, they would put a wooden stick into the stream; the point where it began to smoke was the closest point. By the time the experiments were over, they had determined the electron densities in the exhaust for various concentrations of the additives.

They also learned that if the metallic salts were melted but not completely ionized, they would form droplets, which gave a much bigger reflection than a cloud of only ions. Pratt and Whitney's Willgoos altitude facility in East Hartford had a test cell that could accommodate a radar unit shooting perpendicular to the exhaust plume. It was there that the effects of ions vs droplets were studied in detail. SEI's Brint Ferguson and Joe Klein brought the radar set that was used with the bathtub in the Binney Street lab and made the measurements. In the end, it was determined that a relatively small amount of the material had to be added to the fuel to achieve the desired effect.

On 16 November Johnson wrote,

> I think we have been fairly successful, in that a series of tests has now been instigated installing antennas in the afterburner to see whether we can ionize the gas and essentially provide a faired-over tail cone. Spending a great deal of time myself going over all aircraft systems, trying to add some simplicity and reliability [165].

Feasibility investigations would continue for the next three months. On 15 February 1960, Pratt and Whitney submitted a formal proposal for the remaining development. Over the next five months, they would continue tests on a J57 engine, including evaluating how altitude affected radar attenuation. By 30 July, they would have designed the complete carrier and injection system for the additive. On 25 July, Parangosky recommended to Bissell that the proposal be accepted and additional funds allocated to cover the new work [166].

Fig. 118 $\frac{1}{8}$-scale A-12 RCS model with simulated exhaust, on the "Old Bag." (Courtesy of Lockheed Martin.)

The final choice was a metallic salt that, when vaporized, had a very low ionization potential and was thus very efficient at providing a cloud of plasma. This was simulated for RCS measurements by attaching cylinders with conical tips behind each ejector of a small-scale model and later of the full-scale RCS model (Figs. 118 and 119). The cylinders were actually a wood frame covered with an absorptive material designed to match the characteristics of the ionized exhaust.

Progress

The situation had changed enormously in the year since the Land Panel had ruled against Lockheed's A-3 in favor of Convair's FISH. Johnson wrote that "November has gone by very rapidly, but I think we have made considerable progress." Finally, the full-scale model tests were finally moving along smoothly, and "I think a fine job has been done by all in getting the base ready and the model up and tests underway." Besides the model work at the Ranch,

Fig. 119 Full-scale A-12 RCS model was also fitted with cylindrical frameworks covered with reflective material. (Courtesy of Lockheed Martin.)

the tests which have been undertaken in Florida appear to be very interesting in providing a solution to one of our major problems—that of the rear aspect of the aircraft. I expect to see Frank December 1st and talk over the necessity or desirability of asking for the F-94C for pole tests to either supplement or confirm what has been learned down in Florida.

Anechoic chamber tests back in Burbank had made strides in proving the resistive honeycomb: "the box tests [indicate] that we can load fiberglass honeycomb as effectively as we can the plastic foam. This is an important step in a solution to the structural and high temperature problem." Lockheed had gotten bids from a subcontractor to build plastic leading-edge models to Lockheed's design, and visits to firms in the East were planned.

Johnson had been successful with bringing in engineers from elsewhere in Lockheed. One was working on ferrites, another on anechoic chamber measurements, and two others would supplement the team at the Ranch.

He also reported that "the $\frac{1}{8}$-scale model has been brought up to the latest aerodynamic configuration and is available for tests on the bag. The 10 foot fuselage section is available for test on the large pole, and about half of the eight different configurations of chines are available for tests." Because the $\frac{1}{8}$-scale model, the 10-ft fuselage section, and some full-scale model sections could only be modified in Burbank, Johnson suggested obtaining a C-130B to fly material and personnel between Burbank and the Ranch.

Besides the antiradar work, great strides had been made in the overall design, with structural design proceeding for the entire airframe. Tooling had started for the nose section and equipment bay; the bay had required a redesign to allow it to be pressurized, should the equipment require it. The landing gear had received intensive study, and water-cooled brakes were investigated as a way to reduce the overall weight.

One good piece of news was that a titanium vendor had promised that prices would drop substantially from early estimates as soon as a large order was placed. This savings would help cover other items that were now expected to be very expensive. High-temperature bearings, for example, were being quoted at 20 to 40 times the cost of normal ones. And it was at this time that an order for a high-temperature hydraulic fluid resulted in the delivery of a sack of white powder that would become a fluid once heated to at least 200° F.

One major change in responsibilities was that Lockheed would design and build the ejector, which had originally been assigned to Pratt and Whitney. P&W had not been able to seriously investigate potential flutter problems in the augmentor and tail flaps.

Johnson reported that the low-speed wind-tunnel tests had been completed, and the high-speed model had been shipped to NASA Ames Research Center for tests in the unitary wind tunnel.

Planning was also proceeding on a test rig for the complete fuel system to be used in Burbank. This would hold all of the tanks, piping, and pumps in their correct relative locations and tilt the assembly at angles of at least 30 deg to be sure that fuel would always flow as expected. To prevent this from sticking up above a revetment and being visible across the airport, the test rig would sit in a 70-ft-long, 25-ft-deep pit that had once been used for liquid-hydrogen fuel tests for the CL-400. Because the tests would also include high-altitude simulation, the pit would help contain debris if a tank were to burst.

Another piece of construction in Burbank would be a 20 × 50-ft insulated building for testing a complete aircraft forebody—including the cockpit and equipment bay—at temperatures up to 800° F [167].

December saw an increase in the pace of work at the Ranch. The Agency granted EG&G approval to work 48-hour weeks through 15 January 1960. EG&G was authorized to spend up to $28,000 on construction of the improved pole shield, including $16,000 that had already been paid to subcontractor Ward and Ward [168]. The new shield looked superficially similar to the wing-like structure with an open back that Convair had built for use at Indian Springs. Wayne Pendleton, who joined EG&G late in 1960, recalled that

> It was tapered so that the top was narrow equal to the rotator bottom diameter and the bottom was wide. It was a modified cone structure with the hinged leading edge looking like the leading edge of a wing. The taper helped reduce the cylindrical pole backscatter to the radars and reduced model/shield interactions. After the pole was erected the crew

> would close the shield around the pole. (Pendleton, Wayne E., e-mail to author, 21 Feb. 2008.)

High-speed wind tunnel tests at NASA were very promising, with the design meeting or exceeding the basic performance and stability estimates. An inlet duct model was under construction, as was a model to be instrumented for a temperature survey.

Two problems with titanium first appeared at this stage of development. The first was lack of availability, which was slowing construction of the forebody. The second was machining the hard metal. Johnson reported to Bissell that

> We are having good luck making the various pieces, but the tooling required is considerably more sophisticated than we have previously used. We are having to heat the tooling to bend the material properly. Installation of our special heat treat furnaces and various treating tanks is proceeding rapidly, with completion expected about the middle of January.

To avoid oxidation of the titanium during treatment, it was coated to exclude air. This avoided having to buy extremely expensive furnaces that would heat the pieces in an argon atmosphere.

During December, the pole rotator had failed, and lower priority tests had been moved up so that progress would continue during the repairs. The repairs also improved the rate of rotation and the speed of raising and lowering the model, so that the overall test rate was almost tripled. The exhaust ionization had proven successful, and Johnson wrote that "It appears that the very difficult problem of suppressing the return from the aft section of the aircraft has been successfully solved over the required frequency range."

Johnson was also able to confirm that Lockheed, not P&W, would build the ejectors, which they could do at a slightly lower price. P&W, on the other hand, would assume responsibility for the accessory gear box and drive shaft, which would be attached to the J58 engine. That arrangement would eliminate Lockheed's having to transport the prototype equipment to Florida to fit it onto the engine and get it to work.

The rate of obtaining security clearances was also starting to constrain work. One example was the air conditioning system because the subcontractor had not been cleared for their part of OXCART.

Finally, the hope to extend the A-12's speed from Mach 3.2 to 3.5 was running into problems, mostly related to materials problems at the higher speed:

> Wing tank sealants, electrical wiring insulation, canopy rubber seals, etc., will pose major problems in terms of obtaining satisfactory service life. As we have discussed before, I think that the bookkeeping that it [sic] will be necessary to have to record times on various components

and overhaul procedures will be far beyond anything we have faced to date. A single base operation certainly will do much to simplify problems associated with the above [169].

Full Go-Ahead

On 20 and 21 January 1960, Bissell and his advisors conducted a review of the progress on the A-12. In addition to an Air Force representative, the group included Edwin Land, Guyford Stever, and Edward Purcell. DCI Allen Dulles summarized the results in a memorandum officially notifying the Director of the Budget of the decision. Dulles reviewed the approval given by the President on 20 July and that continuing the project had been contingent upon validation of the RCS and performance objectives that at the time seemed achievable. Dulles concluded that after six months' more work, "The result was favorable in that the evidence presented, which was far more definitive than that available at the time of the original decision, strongly supported the conclusion that the objectives discussed with the President could be substantially achieved. Accordingly, the Agency is proceeding with the Project" [170].

Bissell had given Johnson the word on 26 January. Kelly later wrote, "Talked to the Director of the Program Office. He told us we had the project. We are not sure whether it is 10 airplanes plus a static test, or 12 airplanes plus a static test, but we are in!!" On 30 January, he was notified that the Agency would procure 12 aircraft.

In his progress report for January, Johnson said that some RCS investigation was still needed to firm up the design of the inlets, the lateral fairing of the ejectors, the leading-and trailing-edge detail construction, and the filler material for the chines. The uncertainties of the edge construction and of the RAM for the chines were the source of the greatest risk in meeting the weight budget. Otherwise, there was enough information available to begin construction of about 90% of the aircraft.

The last 17 engineers were in the process of being cleared; Johnson expected that clearances for engineering personnel would then be complete, except for flight-test engineers, who would not be needed for months [171].

Design Revision

The design evolved further after analysis of the original full-scale RCS model. The fixed vertical stabilizers with their swept-back trailing edge were replaced with all-moving rudders with the trailing edges swept forward, similar to the change in the final KINGFISH model. As with KINGFISH, the purpose of the forward sweep was probably to use the wing structure to shield the rudders from radar. Increasing the size of the rudders gave two-and-a-half times more control authority in the thin air at 90,000 ft [129].

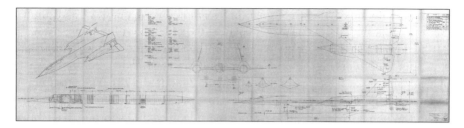

Fig. 120   A-12 final design general arrangement drawing, by Herb Nystrom. (Courtesy of the family of Edward P. Baldwin.)

The other obvious change was the reduction in the size of the boat tail. In the original design, the trailing edge of the inner section of the wing was a straight line from the edge of the afterburner to the tip of the fuselage. By the latter half of October 1959, wind-tunnel testing had shown longitudinal stability problems caused by the chines. To compensate, the trailing edge with the elevons was brought back from the afterburner at a sharp angle before turning toward the tail. The increased elevon authority corrected the problem. The revised general arrangement drawing was completed by Herb Nystrom on 8 February (Fig. 120).

By 18 February, the design changes had been incorporated in the full-scale RCS model, and it was returned to the pole (Fig. 121). EG&G personnel worked overtime and holidays through February and March 1960 to get the

Fig. 121   Full-scale model of late A-12 configuration. (Courtesy of Lockheed Martin.)

measurements done. In early April, high winds during the days had limited the times that the model could be raised on the pole; the solution was to work during low-wind times in the early morning and late evening. This meant expanding the working hours from a 0600 hrs start and 2130 hrs finish to instead starting at 0300 hrs and finishing at 2400 hrs [172]. The wind was a serious problem; at its worst it could cause small pebbles to roll along the ground. When it struck the model from the side, it caused a torque that tended to turn the model into the wind. As a safety measure, the pole mounting had a shear pin that would snap and allow the model to turn; facing the wind caused the least force and the least risk of damaging the model or the pole.

Sixty-five-hour work weeks continued during the summer through the end of October [173]. By that time, it had been necessary to refurbish the metallic foil covering the full-scale model, which had sustained wear and tear during its 11 months of use, despite being taken down and stored in a hangar when not in use.

The expansion of radar facilities continued, with the approval in September 1960 of two 16-ft-diam dishes for 1400-MHz measurements [174], and of a 45-ft dish in 1961 for in-flight measurements [175].

The full-scale model lived on after the A-12 work was completed. When work began on the Air Force's SR-71, the front of the model forward of station 715 was replaced with a section designed to match the new aircraft. Eventually, when all of the work was completed, the model was burned, and the remains were buried (Interview with Ed Lovick, 8 March 2008).

THE PDP-3

As the amount of radar testing at the Ranch increased, the scientists at SEI decided that the data had become too much to process by hand. To speed things up, they decided that a computer would be needed. The difficulty was that no computer within their budget was available off the shelf. So they decided to build their own.

In October 1960, the Digital Equipment Corporation (DEC) had produced a specification for the PDP-3, a new system in their line of programmable data processor minicomputers. Ed Rawson took charge of the project and with the help of Chuck Corderman and Jay Lawson designed and built a PDP-3 using standard DEC logic modules. [EG&G personnel sometimes teased Rawson that the SEI folks must have held stock in DEC. (Pendleton, Wayne E., e-mail to author, 14 Feb. 2007)] Because disk drives were not available, a tape loop running through an Ampex tape drive held intermediate results; eventually, the tape loop was replaced with a drum memory. The project was run like a homebrew computer project, with more emphasis on getting the machine and software to run rather than on making it well documented and easy to use. The design evolved so rapidly that when one of the engineers returned after a two-week absence, he didn't recognize it (Interview with Daniel Schwarzkopf, Stow, MA, 30 Nov. 2003). The design evolved away from the original PDP-3 architecture, and it came to be called CASINO, for computer able to select internal orders.

Eventually the system worked. Radar data were recorded by EG&G at the Ranch on 1-in.-wide data tapes and shipped to SEI in Waltham, Massachusetts. The data could be processed correctly, but the computer could usually only be operated with Rawson looking over the user's shoulder. Eventually the PDP-3 was discarded; one computer engineering textbook stated that in the early 1980s it was running somewhere in Washington state, but the author of that book could not confirm it (Bell, Gordon, e-mail to author, 13 Feb. 2007). There is an unconfirmed report that it was donated to a Boy Scout troop and eventually given to Dow Chemical for disposal. It was the only example of a PDP-3 ever built.

LATE CHANGES

There were only a few structural modifications to the A-12 once manufacturing began (Fig. 122). One was to the wing insertion procedure. The initial technique of splicing the wing beams to the fuselage rings took too long. Later aircraft were assembled by splicing the wings inside the fuselage.

Aircraft are notorious for gaining weight through their lifetimes, and the A-12 was no exception. A second structural change was beefing up the joint at fuselage station 715 to accommodate the increased weight of installed equipment.

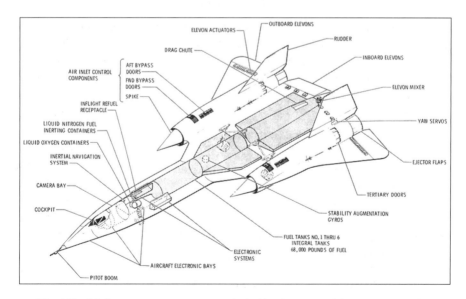

Fig. 122   Major systems arrangement in A-12. (Courtesy of Lockheed Martin.)

The third change came in early March 1961 after wind-tunnel testing showed that the lift developed by the outboard section of the wings—beyond the engine nacelles—would cause excessive loading on the nacelle ring carry-through structure. The solution was to reduce the lift of the outboard wing by adding conical camber to the outboard leading edges (Interview with Sam Kelder, Cottage Grove, OR, 12 Jan. 2003). One way of fixing the problem would have been to redesign the entire wing, which would have caused an unacceptable delay and the scrapping of large assemblies. By fixing the outer part of the wing, the work could continue, and the schedule hit was minimized.

The design of the composite rudders included an internal air space to reduce their weight. After some of the early heat-soak flights, the rudders were found to have lost pieces of material. The problem turned out to be that the air trapped in the cavity in the rudder had expanded and cracked the material. The fix turned out to be simple: drill a $\frac{1}{8}$th-in. hole to vent the cavity. This allowed the air to escape as the aircraft climbed and heated and to return as it descended (Interview with James D. Eastham, Rancho Palos Verdes, CA, 21 Dec. 2002).

Procuring asbestos for the radar-absorbent components of the skin proved to be a challenge. By the time the aircraft was in production, the Occupational Safety and Health Administration (OSHA) had banned most uses of asbestos. Lockheed had ordered some from John Mansville. Ed Baldwin remembered that they

> came back to us and said it's a banned substance and may not be used on any new design. ... We said, "Well, we really need it but we can't tell you what we're doing." They said, "Well, we can't sell it to you."
>
> Our purchasing man got hold of the guy back at the Agency who was in charge of things, and he said, "If you need asbestos we'll get asbestos for you. Just give us all the information on what you need, the quantities you need, in what form, how many pounds you need a month, and so forth, and we'll get it for you." And they did. (Baldwin, E. P., quoted in Baldwin, R. E., unpublished transcript of tour of SR-71 by Ed Baldwin and Lou Wilson, Beale Air Force Base, California, 30 Aug. 1989.)

The asbestos required special handling. Anyone filing or grinding it had to wear respiratory protection and use a special vacuum cleaner to capture the dust into special plastic bags. Baldwin explained that,

> Today there are materials you could use instead, high strength high temperature carbon fibers, silicone resins, and the like that you can bake and hot press and get good results. There was none of that in those days. We knew the asbestos would do what we wanted it to do and it is used on the plastic part of the chines. (Baldwin, E. P., quoted in Baldwin, R. E., unpublished transcript of tour of SR-71 by Ed Baldwin and Lou Wilson, Beale Air Force Base, California, 30 Aug. 1989.)

THE DISH

In early 1961, EG&G began to build up the radar systems at the Ranch for flight testing of the A-12. This meant doing what an enemy radar network would do: acquiring the aircraft as far out as possible and then illuminating it with a fire control radar to get accurate measurements. Moreover, they had to determine the angles from which the A-12 was being illuminated at any instant so that they could know how large the reflections were in every direction. And to avoid wasting test time, it was necessary to acquire the aircraft and point the fire control radar in the shortest possible time. To accomplish that, the A-12 did something that an aircraft on a mission would never do: transmit signals to help aim the radar.

The acquisition radar used a 60-ft-diam dish antenna, which was built by the D. S. Kennedy Company, in Cohassett, Massachusetts, the builder of the antennas for the Distant Early Warning (DEW) Line (Fig. 123). At first, a set of yagi antennas on the edge of the dish detected the A-12's telemetry signal and was used to drive the motors that steered the dish. When the dish was approximately aimed, control was transferred to a set of dipole antennas at the focus of the dish; with the much higher gain of the dish, this provided much more accurate tracking information than the yagis. With the help from

Fig. 123   EG&G facilities at the edge of the lakebed at the Ranch included the 60-ft dish antenna. (Courtesy of T. D. Barnes.)

the telemetry signal, the A-12 could be acquired as far out as 300 miles, effectively as it came over the horizon.

The fire control radar was an X-band system designed to guide Nike-family surface-to-air missiles. The Nike radar used a very narrow beam, and using it alone to acquire the A-12 would have been difficult or impossible. Until the dish's dipole receiver acquired the A-12, the Nike radar was slaved to the dish, pointing in the same direction as the dish. This help from the dish meant that the Nike radar could acquire the aircraft very quickly. Control was then switched to the Nike radar, and its signals aimed both its antenna as well as the dish (Schwarzkopf, D., e-mail to author, 16 April 2008). This slaving of the two radars was something that could never have been done in an operational environment, where the antennas are not located together, and where the operator of the acquisition radar would have to transmit the aircraft's location to the operation of the fire control radar.

The telemetry system to report the A-12's attitude was designed by SEI's Brint Ferguson and built by EG&G's Wayne Pendleton. The airborne component used gyroscopes and other attitude-indicating instruments that modulated a 400-MHz transmitted signal. The ground component included an analog computer that converted the received signal into coordinates that indicated the A-12's attitude in three dimensions.

The range and elevation information from the Nike radar, combined with the attitude information from the telemetry system, gave the exact angles at

Fig. 124 A-12s and YF-12As on north ramp, with RCS measurement facility buildings in background. (Courtesy of Lockheed Martin.)

which the A-12 was being illuminated when every reflection was measured. The data were all recorded on magnetic tape for later processing, but they were also plotted on paper as it was taken, so that the team would know immediately if there was a problem with the signals, and the test could be rerun without delay (Fig. 124).

Things did not always go smoothly. When the signals from the radars came out 10 dB lower than expected, Jay Lawson came out from Massachusetts to help debug the problem. Eventually he located the problem: Pendleton had put in an attenuator in the signal path and had failed to account for it in his calculations. The hardware did not have to change, but the equations did (Interview with W. E. Pendleton, Torrance, CA, 8 Oct. 2005).

The radar systems would remain in operation until 1967, when work on the A-12s was essentially complete. In the mid-1970s, they would be reopened for work on the HAVE BLUE stealth technology demonstrator and on its successor, the F-117.

Chapter 15

PROPULSION

The engine nacelle and the J58 engine in it were the keys to the A-12's extraordinary performance. The nacelle was designed primarily by David Campbell, a member of Ben Rich's thermodynamics team. The problem that the nacelle had to solve was matching the airflow from the inlet to that needed by the engine at just the right speed, temperature, and pressure. Campbell's solution was a divergent-convergent inlet containing a movable centerbody with a conical tip (a "spike") to position the shock wave and various vents to remove excess air (see Fig. 125).

The first step was to slow the Mach 3 air at the inlet to subsonic speed. As the air moved through the inlet, it would generate various shock waves. The first one, off the tip of the spike, was conical in shape and was known as the oblique shock.

The initial section of the inlet narrowed as the air moved away—diverged—from the centerline of the inlet. It contained various shock waves as the air slowed. The narrowest part was called the throat and was the location of the final shock, called the normal shock, which was perpendicular to the airflow; behind this "normal" shock the airflow was finally subsonic. Behind the throat the inlet grew wider and as the airflow converged back toward the centerline it continued to slow until it reached the face of the engine. This area was called the diffuser.

To control the internal pressures and position the normal shock at the throat, air was bled from the inlet in two directions. At the widest point of the centerbody, its surface was slotted, allowing air to flow into it (Fig. 126). From the interior of the centerbody, the air flowed out through the four support struts and exited overboard. At the throat, air vented into a space in the outer circumference of the nacelle, called the shock trap. From here it was directed back and along the outside of the engine to cool it. During ground and low-speed operation, this cooling air was supplemented with air from "suck-in" doors on the nacelle.

If there was too little airflow, combustion would be inefficient, and the engine would not develop the necessary power. However, if there were too much airflow, the excess air would have to be dumped overboard, which caused drag and reduced the range of the aircraft. And if the backpressure

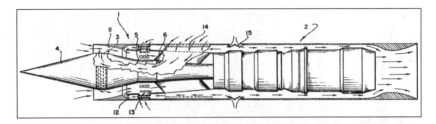

Fig. 125 Lockheed's David Campbell posthumously received patent 3,477,455 for his design of the A-12's axisymmetric inlet. (Courtesy of the U.S. Patent and Trademark Office.)

grew too great, the normal shock would be expelled from the inlet, and air would spill out, creating a bow wave around the lip of the inlet. This was called an unstart, and it would cause a huge loss of thrust and increase in drag, violently yawing the aircraft toward the side with the unstart. The loss of cooling air from the shock trap would also cause the engine to overheat.

From 1959 into 1961, the inlet was tested in the NASA Ames Research Center unitary plan wind tunnel, which could achieve speeds of Mach 3.5 (Fig. 127). Before each test, a test plan would be written up, describing the purpose of the test, the speeds, the positions of the model relative to the wind, the locations on the model where measurement probes would be located, and the formulas to be used for data reduction.

The inlet portion of the model was typically constructed of steel, whereas the rest of the airplane body was aluminum alloy. Temperature-sensitive paint, which would change color wherever the model's temperature exceeded a certain threshold, was sometimes used.

The purpose of each run varied depending upon what design options were being studied. For a series of runs beginning 13 June 1960, various spikes in two different sizes were used to give 5% and 10% larger throat areas than

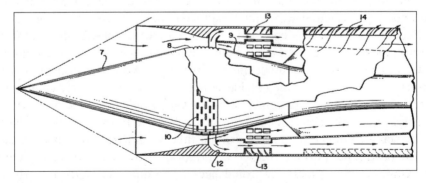

Fig. 126 Inlet with oblique shock and internal air flows. (Courtesy of the U.S. Patent and Trademark Office.)

Fig. 127 A-12 with model 204 inlet. (Courtesy of NASA.)

in an earlier test series. The spikes had different amounts of bleed and different configurations of bleed vents [176].

By January 1961, the sizes of the spikes and the inlet throat apparently had been decided because the tests run that month were to measure the accuracies of the spike actuator and of the main control and bypass actuator, as well as to evaluate changes to the inlet, including a system for expelling centerbody bleed air through louvers in the nacelle, and plenums and ducts for metering cowl shock trap bleed air and bypass air [177].

Eight months later, the centerbody bleed was still being investigated. In September 1961, the tests focused on modifications to the centerbody bleed system, measurement of internal cowl static pressures, and determination of the optimum configuration for the centerbody porous bleed area [178].

Because the A-12 would cruise in a slight nose-up attitude (i.e., with a positive angle of attack), the nacelle was attached to the wing pointing slightly down (1 deg 12 min) below the line of the wing, which it was hoped would point it directly into the relative wind. As wind-tunnel studies continued while the prototype was assembled, it was discovered that the airflow at the inlet was not from directly ahead, but was moving outward from the nose of the aircraft toward the side. Moreover, the angle of attack at cruise was higher than expected, and so the airflow was moving upward relative to the angle of the wings even more than had been expected.

The solution was to modify the inlet to better align it and the spike with the direction of the air. The spike was tipped downward and inward. Fortunately the fix was made early enough that no major parts of the nacelle on the prototype had to be scrapped.

The J58 Engine

Although Kelly Johnson's earliest Mach 3 design in April 1958 had used the Pratt and Whitney J58, the engine that eventually flew in the A-12 in 1963 was very different from the design that had existed five years before. It was the product of a long and painful evolution.

Pratt and Whitney's first Mach 3 engine was designed for the XB-70 Valkyrie bomber. Its earliest incarnation was as the FX-114 prototype, and later versions were named the JT9. Bob Abernethy recalled, "I was shocked when I saw the JT9. It was so big! It must have been the largest engine in the world at that time. You could almost stand up in the afterburner" [179].

When General Electric won the XB-70 contract with their J93, Pratt and Whitney took aim at the Vought F8U-3 Crusader III fighter. The East Hartford development center downsized the JT9 by about $\frac{1}{3}$, and the JT11 was the result. It was a conventional afterburning turbojet with 26,000 lb thrust. The Navy designated the engine name as the J58. (Jet engines for the Navy traditionally used even numbers, and those for the Air Force used odd numbers.) The prototype engine tested for the Crusader III was given the designation J58-P2 (Fig. 128). It was significantly different from the later J58-P4 recover bleed air engine.

Early in the J58 project, P&W had about 25 engineers assigned to it. Bill Brown, P&W's chief engineer, was overall responsible. Other early key

Fig. 128   The J58-P2 was an early prototype. (Courtesy of Roadrunners Internationale.)

personnel were Don Pascal, Norm Cotter, Dick Coar, Ed Esmeier, and William Gordon [180]. Eventually Pratt and Whitney had a total of 2000 working on design, testing, and production.

One of the numerous experiments carried out at the company's West Palm Beach, Florida, facility was the use of high-energy fuel (HEF), also known as "zip." Because the fuels contained boron compounds, they were highly toxic. It was hoped that the increased engine thrust would be worth the special handling needed.

Unfortunately, it was not to be. A new problem arose, plugging of the fuel system and afterburner spray bars. The accumulation of material even increased the friction on the turbine. With normal fuel, the engine would continue to spin for several minutes after fuel was shut off. After using HEF, the engine would grind to a halt in just 30 seconds. And the toxicity of even the burned fuel was proven when vegetation died in a 2000-ft-long swath behind the test stands.

DESIGNING FOR A NEW MISSION

After the Navy chose a different engine for the F8U-3, the CIA was the only remaining customer for the J58. Adapting the engine from its original application in a Navy aircraft to a new one in a strategic reconnaissance aircraft meant a fundamental change in the priorities for its operation. If a naval aircraft engine suffers a partial failure, it should still be able to operate well enough to allow the aircraft to recover back onto the carrier, or at least to fly away from the carrier and allow the crew to eject. This calls for operation at low altitude and low speed.

The top priority for a strategic reconnaissance aircraft, on the other hand, is to exit denied territory so that the aircraft and crew cannot be captured. This means that a partially functional engine must continue to operate at high altitude and high speed. In the event of a major failure, the engine control will set the fuel control, inlet guide vanes, and all of the other engine actuators to a predefined state. In the redesign of the J58, these had to be set for operation at Mach 3.2.

Although Pratt and Whitney had experimented in the mid-1950s with engine controls using vacuum tube electronics, no practical systems had resulted. When the J58 was being designed, mechanical linkages were used to connect the control unit to the individual engine actuators. Typically, the control would drive a number of rods, which rotated with changes in temperature and translated (moved parallel to the length of the rod) with changes in pressure. At the end of each rod was a three-dimensional cam, and a rod called a follower rode on the surface of the cam and moved up and down with changes in its shape. The follower drove a cable to the actuator.

For operation in case of a failure, the engine control would set the drive rod all the way in or out and turn it all of the way to a stop. The surface of the cam at the place the follower then made contact was known as a plateau; the height of the plateau from the centerline determined the setting of that engine actuator in the emergency state. For the J58 used in the A-12, the plateaus drove the actuators to the high-speed high-altitude settings.

The redesign that led to what was called the JT11D / J58-P4 turned out to be so extensive that it was referred to as "laying down a new centerline." In principle, the change was so large that it could have been considered a new engine project. However, treating it as a redesign within the scope of the original project avoided a great deal of bureaucratic overhead. On the other hand, it also masked the difficulty of what was being done and led to questions about why progress was not faster.

The main problem that had to be solved in the redesign of the J58 was heating of the air entering the engine. The purpose of the compressor of a turbojet is to increase the density of the air entering the engine; without this compression, the amount of air in the combustion chamber would not provide enough oxygen at high altitudes to burn enough fuel to provide sufficient thrust. At high Mach numbers, the "ram air" compression caused by the air's motion through the inlet heats the incoming air and limits its compression so that not enough air is available for combustion.

Another problem is that the faster the air enters the compressor, the faster the blades in the compressor disk must turn to avoid stalling, which would cause the blades to flutter, a source of stress leading to metal fatigue. However, high rotational speeds put more stress on the compressor disks. Having no compressor, a ramjet avoids these problems, but it does not work at low speeds, which is why several of the early Lockheed designs had both turbojets and ramjets.

In a talk at a Pratt and Whitney reunion, Bob Abernethy recalled,

> In October 1958 the solution for all these J58 problems was clear to me. Bypass the bleed air around the compressor at high Mach number into the afterburner and increase the mass flow and thrust significantly. Actually it converted the engine into a partial ramjet with capability above Mach 3! [179].

It took Abernethy six months to work out all of the details and convince his manager, Norm Cotter, that the design would work. The final piece fell into place in April 1959 when he convinced Cotter that "... if we opened the bleeds at Mach 2.2 there would be no flow increase, no bleedflow, no delta P from compressor mid stage to the afterburner. And that is what we did. It was an absolutely smooth transition as I predicted" (Abernethy, R. A., e-mail to author, 4 Feb. 2008). Cotter took the idea to Bill Brown, who called Kelly Johnson on the encrypted telephone they used for sensitive communications. Johnson apparently checked with his thermodynamics team and then approved the idea. At last Abernethy was allowed to apply for a patent.

On his patent application, Abernethy called it the recover bleed air engine (Fig. 129), and the technique became know as "bleed-bypass." (At least

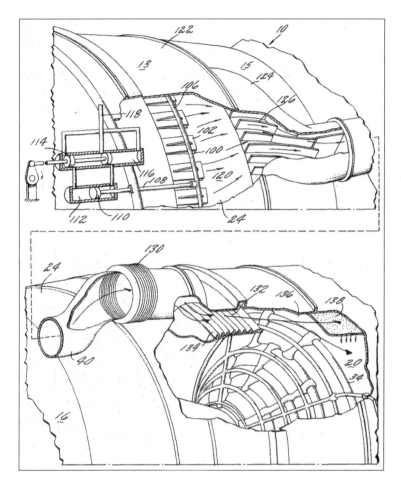

Fig. 129 J58 air bleed (upper drawing) and reinsertion (lower drawing). (Courtesy of the U.S. Patent and Trademark Office.)

one manager made a snide remark when he realized that the initials of the engine—RBA—were the same as Abernethy's.) Six tubes were used to take bleed air from the compressor, bypass the engine core, and reinsert it into the afterburner. This allowed the J58 to operate as a turbojet at lower Mach numbers with the compressor increasing the density of the incoming air enough to support efficient combustion and to operate as a ramjet at higher Mach numbers, letting the inlet compress the air. In bypass mode, the engine core would only receive enough air to keep it operating at a safe temperature.

More Changes

Another of Abernethy's suggestions for engine control was to use variable inlet guide vanes. The inlet guide vanes (IGVs) are located in front of the first compressor stage; they are fixed and do not rotate like the compressor blades behind them. However, their pitch can be changed, as a constant-speed propeller varies the pitch of its blades. The purpose is to give better control over airflow and to keep the compressor blades behind the stator from fluttering. Abernethy was chastised for using a General Electric design technique. (Pratt and Whitney traditionally used the bleed technique of venting air from the engine core.) Nevertheless, the variable IGVs were eventually incorporated in the final ("K") model of the J58 (Fig. 130).

Abernethy had not been cleared for Projects GUSTO or OXCART. For his analysis of the J58's performance, he chose flight conditions of Mach 3.2 and 90,000 ft, which—unbeknownst to him—were the target for cruise performance of the follow-on vehicle. Abernethy was transferred to work on the RL-10 rocket engine and always wondered whether the transfer was because he knew too much about the project.

Besides the bypass ducts, there were several other changes between the Navy and CIA versions of the J58. One was adding a ninth compressor stage. Another was adding larger air bleeds from the engine.

The spacers between the compressor disks were given holes to allow cooling air to circulate. In one earlier engine run, the compressor had overheated.

Fig. 130   This late-model J58 shows three of the six bypass tubes. (Courtesy of Roadrunners Internationale.)

When it was cooled off and disassembled, one of the disks had been so damaged by the heat that it had become flexible, and even after it cooled it could be deformed with finger pressure. It became known as the "flubber disk," after a magical substance from the 1961 movie *The Absent Minded Professor*.

Another change turned out in the long run not to be needed. There were two sets of spray bars to inject fuel into two zones in the afterburner. The plan was that fuel flow to each set would be separately controlled. Bob Boyd recalled that

> testing showed that separate control and separate pipes was not necessary and the control was simplified to a single system. But the simplification was simply to remove the second set of controls and not to completely redesign the A/B fuel control. The plumbing was simply replaced with a "Y" joint from the single source to feed the two zones. (Boyd, B., e-mail to author, 13 Feb. 2008.)

An innovation—in the realm of turbojet engine controls—was the use of fuel as the hydraulic fluid. In fact, fuel was used for hydraulic fluid not just for the engine controls but also for the aircraft flight controls.

The final change from the original J58 was that the engine control built by Hamilton Standard was replaced by one built by Bendix for General Electric's J93 engine. This happened despite the fact that Hamilton Standard was owned by United Technologies—Pratt and Whitney's parent corporation. It was another case in which the urgent need for success overrode the not invented here (NIH) syndrome. In this case it happened very quickly. According to Pratt and Whitney lore, it was found that one of the Bendix controls was sitting on a desk in the Naval Air Systems Command (NAVAIR). One night it simply disappeared and then reappeared shortly afterwards at Pratt & Whitney's Florida Research and Development Center (FRDC).

An unusual approach was used in designing the algorithm to be implemented by the engine control. One of the problems was to make sure that the engine would not run too hot, which could melt it. This meant that as fuel flow was increased—which would increase thrust but also increase temperature—care was needed to be sure that the temperature did not become too high, and that if the temperature did go over the limit then it would be brought back down very quickly. The engineer in charge of an engine test stand—known as the conductor—could control the engine with manual controls and had learned how to maintain the temperature.

On the first day the new control system was installed on test stand A1, Tom Warwick bet the conductor a dollar that the automatic system was better than he was. Warwick lost the bet, but watched what the conductor did with the controls. He saw that when temperature was increasing, the conductor increased the fuel flow very slowly; if the temperature went too high, he would quickly turn it back to the minimum setting and then slowly start back up. Warwick changed the control schedule to imitate this, and on the second day he won the bet. He used the betting technique several times to get ideas for improving the control system's performance. He later described the control system as an asymmetric, pulse-modulated, variable gain control that emulated the A1 test stand crew (Telephone interview with T. Warwick, 23 April 2008).

SIMULATIONS

The opening and closing of the bypass doors would be a critical operation every time the aircraft flew at Mach 3. To understand whether it and other aspects of engine operation would work, Pratt and Whitney engineers performed computer simulations of the inlet, engine, and ejector. In the late 1950s and early 1960s, the use of digital computers for simulating physical systems was in its infancy. Although digital simulations were used for steady-state operation, calculation of transient operation—that is, when engine conditions were changing—took much more computation, and simulations detailed enough to be useful would take enormous amounts of time.

On 7 November 1959, a Convair B-58 was destroyed during a test flight. It was suspected that an unstart on one or both engines on the same side of the aircraft had caused the uncontrollable yaw that led to its breakup. In an unstart, a pressure rise inside the engine nacelle would expel the shock wave from the inlet, leading to a rapid decrease in engine thrust. Concerned about what unstart would do to the A-12, Norm Cotter called Warwick into his office and asked him what would happen to the thrust of a J58 engine experiencing an unstart. Warwick asked for an hour of time on the IBM 709 computer. Ordinarily, engineers were not allowed to program the computer, but could only tell the full-time programmers what the program should do. However, this time the answer was needed urgently, and Cotter arranged the time.

Warwick had studied the FORTRAN programming language and worked out a way to use the steady-state analysis program to approximate transient analysis. He dictated the new lines of the program to computer guru Harry Williams, who typed them onto the punched cards used for programs. After an hour of running time, the program gave what appeared to be a reliable answer: If the J58's inlet were to suffer an unstart, within one-tenth of a second the thrust would drop by 50%. This first rough estimate was the beginning of Pratt and Whitney's use of digital simulation for transient analysis.

However, for the next few years analog computers continued to provide the most accurate simulation of transients. Where a digital computer represents pressures and temperatures as numbers and performs arithmetic on the numbers, an analog computer represents them as voltages; the larger the temperature or pressure, the larger the voltage; the faster the temperature or pressure changes, the faster the voltage changes. Some components of the simulation are actually servos—motors—that move as the voltages change. The outputs of the simulations are typically from strip chart recorders, where motor-driven pens trace wriggling lines on moving strips of paper.

One problem with analog computers is that if the voltages in the simulation get too high, they can actually burn out parts of the computer. Tom Warwick remembers simulating opening and closing the bleed-bypass tubes after the aircraft had already reached Mach 3. The pens of the strip chart recorders would vibrate so rapidly that they would spatter ink. And occasionally servos burned out, a significant expense.

What the simulations meant was that opening or closing the bypass would cause a transient change in nacelle pressure of as much as 20%. That pressure change was enough to expel the shock waves from the inlet and to cause pressures throughout the engine nacelle to oscillate wildly.

The unstart problem was not just a computer problem; it meant trouble for any real-world aircraft using the engine and nacelle. At that time, Warwick had no idea whether the engine was being developed on spec as a research and development (R&D) exercise, or whether it was intended for a real aircraft. He went to his boss, Norm Cotter, with a warning. He explained the simulation results and said that if the engine was for use in a real airplane, then an unstart could be violent enough to threaten the aircraft and the pilot and that "someone could die." Cotter listened carefully and said simply, "You're excused." Within a few weeks, Warwick had been cleared for more access to Project OXCART and learned that the engine and the aircraft were for real (Telephone interview with T. Warwick, 18 June 2004).

Bob Boyd had joined the J58 project straight out of college in 1959. Like many of the engineers at West Palm Beach, he was young and naïve. They worked hard on the project and didn't worry about asking who was funding it or what aircraft it would fly in. It was three years before he learned. In 1962, he accompanied a JT11D-20 to East Hartford for a shakedown of a test cell. There the more experienced engineers started asking what the engine was for, and he had no idea. The only answer he got was during a dinner with a senior manager, George Armbruster, who hinted that "something big" was coming (Telephone interview with B. Boyd, 9 Feb. 2008).

In the summer of 1962, Boyd was cleared, and at the end of the year he accompanied the first J58s to the Ranch. He was one of many P&W engineers and technicians who rotated between Florida and the West Coast. At any one time, there would typically be two engineers and two engineering aides working at the Skunk Works on data reduction. They came to California for six-month assignments and were allowed to bring their families. At one point the Agency became worried about a security breach, when several housewives all from West Palm Beach began appearing as contestants on Los Angeles television game shows and winning. One wife won $50,000, and her husband quit Pratt and Whitney (Telephone interview with T. Warwick, 18 June 2004).

Flight Test

The J58 was not ready in time for the first flight. When the number one A-12 first took to the air on 26 April 1962, it was powered by a pair of J75s. It was not until the end of 1962 that the first J58s arrived at the Ranch and were installed. When the aircraft did not meet the expected performance targets, the Pratt and Whitney team had to work closely with Lockheed to fix it. Norm Cotter, the head of the performance group, remembered,

> One of the most challenging aspect of flight testing when performance is not met, is the simple question of whether the aircraft drag is higher that the wind tunnel scale models predicted or was the engine thrust lower than performance estimated from sea level testing. There is no thrust or drag measuring system in flight, and there is no completely satisfactory way of duplicating Mach 3 in a ground test facilities. In many instances when the type of conflict arise between the aircraft and engine developers, there is a great deal of defensive posturing. This never occurred on this program. We all just went to work to fix the problem without assigning fault. All data was shared openly. At one point (lasting several years) I had a team of my people stationed full time at the Skunk Works. They were headed by Stan Ellis, and included Bob Boyd. They worked hand in hand as part of the Lockheed team. Thinking back, it was this feature of cooperation that made this program very special, and very prized by virtually every person who worked on the program. (Cotter, N., e-mail, 18 May 2008.)

The P&W engineers and technicians assigned to the Ranch and the Skunk Works had to evaluate the performance of the engine and nacelle, diagnose problems, and invent fixes. Their source of data was usually a set of instruments in an equipment bay of the number one A-12, Article 121 (Fig. 131).

By modern standards the equipment was primitive, with a 35-mm movie camera filming a set of instruments. After each flight, the film would be shipped to Burbank and developed, and engineers would then spend hours recording the reading on each instrument in every frame. The raw data would then be shipped to Florida for data reduction and interpretation, in addition to the analysis done by the P&W engineers at the Skunk Works. Evaluation of performance would then be sent back to Lockheed and P&W engineers at the area.

Some engine simulations were performed on an IBM digital computer in the Lockheed "bomber plant" in Burbank, which was more powerful than any other system available to Pratt and Whitney. In more recent times, engine manufacturers have been required to share their computer models of their engines with the customer and with the airframe manufacturer—not so in the early 1960s. Bob Boyd carried the J58 simulation program with him on four reels of tape that no one else was allowed to touch. Pratt and Whitney engineers would use keypunch machines in Burbank to enter the data for

Fig. 131   Article 121 initially flew with J75 engines. (Courtesy of Lockheed Martin.)

each simulation on punched cards. When one of the Skunk Works's reserved time slots on the computer opened up, they would load the program tapes and data cards, run the simulation, collect the output, and erase the program and data from the computer (Boyd, B., e-mail to author, 13 Feb. 2008).

Control of the spike proved to be a major problem in getting the A-12 to operate reliably. For the most efficient operation, the spike had to be positioned such that the shock wave was right at the edge of the inlet. The spike was originally controlled by a pneumatic system built by the Hamilton Standard corporation. Unfortunately, it responded slowly and was difficult to keep calibrated through the enormous range of temperature and pressure changes the aircraft experienced.

In flight, the pressure inside the nacelle would increase, and the normal shock wave would be expelled from the inlet—known as an unstart—causing an immediate drop in thrust and a violent yaw toward that engine. At times the yaw would continue to the point where the other inlet would also unstart and the aircraft would yaw back. The first inlet would restart and push the nose to the other side. The side-to-side shaking would sometimes break the faceplate of the pilot's pressure suit helmet and cut his face.

One of the causes of the pressure variations that led to unstarts was separation of airflow from the walls of the diffuser. Bill Brown suggested to Kelly Johnson that Lockheed add "mice" to the nacelle. Mice were a technique first devised in early jet inlets; a mouse was a bump, usually made of sheet metal, that was riveted to the inside of the inlet to change the airflow. In the case of the A-12's inlet, the mice were so large that they were sometimes called "rats." Whatever their name, they improved the airflow and were an example of the tight cooperation between Lockheed and Pratt and Whitney in getting the engine and nacelle to work together.

Because the early engine control unit was affected by mechanical disturbances, it was constantly changing the engine power even if the pilot had not made a control change. To avoid getting into an unstart or an engine overtemperature condition, the pilot was constantly adjusting the engine fuel flow trim control. It took so much time that he had none left over for navigation or operating the cameras. In an attempt at a quick fix for the problem, Pratt and Whitney devised a way to automate the trim. The plan was to take a signal from the exhaust gas temperature (EGT) display in the control panel and use it to drive the engine trim. Bob Boyd computed a schedule of fuel trim settings as a function of EGT and gave it to Harley Nethkin. He designed a digital control system that took the EGT as an input and gave the trim setting as an output. Although it reduced the pilot's workload, it did not solve the unstart problem (Telephone interview with B. Boyd, 9 Feb. 2008).

This was one time when Kelly Johnson's broad expertise worked against him. Because he was used to having an intuitive understanding of all of the aircraft systems, he was hesitant to replace the pneumatic control system with an electronic one; electronics was the one area where he did not have a deep understanding. Frustrated with the lack of progress in solving the problem, Albert "Bud" Wheelon, the CIA's deputy director for science and technology and successor to Richard Bissell, threatened Johnson with cancellation of OXCART if he would not switch to an electronic inlet control system (Wheelon, A. D., speech to Roadrunners Internationale reunion, Las Vegas, NV, 2 Oct. 2003).

Johnson relented, and Lockheed hired electrical engineer Frederick "Fritz" Schenk. Schenk went to the Pratt and Whitney facility in Florida, where he spent a week with Tom Warwick working on a specification for the control system. Lockheed then contracted with Garrett Corporation to build the system. After it was installed and working, unstarts almost totally ceased.

Chapter 16

NEW COUNTERMEASURES, NEW THREATS

PROJECT KEMPSTER

One active stealth feature was not developed until after the A-12s were flying. Tests showed that a vertically polarized radar wave aimed at the front of the aircraft would couple into the inlets, cause energy to flow along the outside of the nacelles, reflect from the back of the nacelles, travel forward again, and be reradiated back toward the radar unit. Ed Lovick had been concerned with shielding the inlet and came up with a method to accomplish it.

The solution used a technique that—unknown to Lovick at the time—had been conceived in 1956 by Arnold Eldridge, of General Electric (GE), and that received a patent in 1964 (Fig. 132). The idea was to use a particle accelerator to create a cloud of ionized gas around the aircraft and render it invisible to radar. Subatomic particles from the accelerator would strip electrons from the atoms of the passing air, creating a cloud of ions that would absorb the radar energy [181]. This was an application of the same principle that caused a loss of communication with missiles moving at very high speeds and that had been reported by Bissell to the radar team in October 1956. It is uncertain whether Eldridge's work was that which Bissell had said was being funded by the Air Force [182].

Eldridge intended to mask the entire aircraft, which would have taken an enormous amount of power for the accelerator. For the A-12, the technique had to be applied only to the two inlets.

The Agency (CIA) hired Nick Damaskos, a former Boeing engineer who had returned to graduate school, to run what became Project KEMPSTER. The design and construction of the equipment was contracted out to Westinghouse Research Laboratories, where Benjamin LaCroix became the lead engineer. The contract began on 3 June 1963, and by the time the Agency visited on 19 July, an analytical study had been completed, and the contractor was at work on synthesizing possible designs for the equipment. The conceptual design looked very optimistic, with the electron guns weighing between 25 and 50 lb each, and the operational power only 5 kW. At that relatively

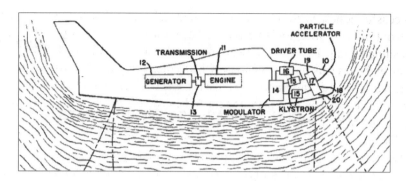

Fig. 132 Use of a particle accelerator to create an ion cloud. (Courtesy of the U.S. Patent and Trademark Office.)

low power consumption, it was thought that a small ram-air turbine could provide the necessary power [183].

The project progressed into experimentation to determine the power of the beam that would be needed to produce a useful amount of radar attenuation (Fig. 133). Initial tests used a vertical electron beam in front of an antenna to measure how much energy was attenuated by the cloud of ions [184].

Eventually the equipment evolved into a test apparatus, which filled the Q bay of the A-12 and sent an electron beam downwards from the fuselage, blocking transmission from an antenna behind the beam. When that proved

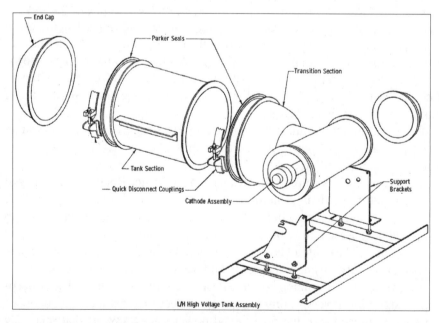

Fig. 133 KEMPSTER high voltage tank for generating electron beam. (Courtesy of the Central Intelligence Agency via the National Archives and Records Administration.)

successful, a new version was built to send a beam sideways from the chine in front of each inlet, ionizing the air before it reached the inlet. The device consumed about 125 kw of power, most of what the A-12 had available. The electron beam was pulsed on and off 100 times per second, causing the radar return to switch at the same rate between the normal high return and a lower amount. The system was installed on A-12 article 126 and was tested first on a pole and by October 1963 in flight, with EG&G operating the radar systems. By December 1964, flights were made using Article 131.

Because the collisions between the high-energy electrons and the air molecules would create X rays, there was concern about how much shielding would have been needed to protect the pilot. Westinghouse conducted shielding studies in a hangar at the Ranch. The studies determined that if lead were used as the shield, the weight would have been prohibitive.

Flight tests showed that every time the electron beam pulsed on, the return from the inlet dropped by an easily measurable amount. That seemed promising enough to proceed with the design of an operational version of the device that could run continuously while taking less power and space. Phase I of KEMPSTER A concluded in March of 1965 with the termination of flight tests, and at the end of June 1965 Westinghouse delivered a final report on the prototype.

Fiscal year 1966 money was allocated in the National Reconnaissance Office (NRO) budget for further work. Part of this was the development of four operational prototype equipment packages as KEMPSTER A Phase II; General Electric was brought into the work at this point and tasked with specifying the electrical power requirements. A new theoretical study by Westinghouse was authorized as KEMPSTER B [185].

Work continued, and on 1 August 1966 contractor personnel made a presentation to the Agency (CIA). Two days later Office for Special Activities (OSA) and Electronic Intelligence (ELINT) personnel, including the deputy director for special activities, John Parangosky, discussed the progress and problems. Cost overruns and schedule slips were such that penalties were consuming most of the contractor's profit. They noted that it was "no small effort" to cut the equipment's weight by half and volume by one-third. As there was no other program of its kind, the Agency was blazing a technological trail that would have applications beyond OXCART. The group recommended that additional funding be provided for KEMPSTER A and that money be obtained from the NRO to fund KEMPSTER B. Unfortunately, the recommendation was rejected by Carl Duckett, who had succeeded Bud Wheelon as deputy director for science and technology (DDS&T) [186].

Some design work was done on the operational system. The mounting system for the package was designed by Donald Bunce and would have fit in the chine on the port side of the aircraft. Bunce was kidded by other engineers about the "Bunce Bump" because the chine would have bulged out to accommodate the package. The Agency named the package CATNAP, and it was one part of an upgrade package to turn the original A-12s into what would have been known as A-12Bs. Because Project OXCART was cancelled, the upgrades never happened.

TALL KING

In late 1959, the Agency's ELINT Staff Office (ESO) began efforts to determine the effectiveness of Soviet radars against both OXCART and U. S. strategic bombers. Of particular concern was a new early warning radar, code-named TALL KING, which was just being deployed. In a 1998 article in *Studies in Intelligence*, Gene Poteat, a former member of ESO, described the various efforts that provided the information needed for the assessment.

The first question was where the radars were located. If a TALL KING transmission were reflected by an object at a known position to a receiver at a known location, then the analysts could solve the necessary spherical trigonometry equations to pinpoint the source of the signal. A receiver tuned to the TALL KING frequency was placed on a 60-ft RCA radar antenna in New Jersey. The reflector was the moon. Lincoln Labs was involved in characterizing the moon's surface and provided information on its reflective qualities that were used in optimizing the receiver. As the moon revolved around the Earth, transmissions from anywhere in the Soviet Union could be captured. The study showed that there were a large number of TALL KING installations, and they provided "incredibly complete" coverage.

The next question was the precise power radiated by a TALL KING. Contractors developed a series of power and pattern measurement systems (PPMS) that were flown in various aircraft around the periphery of the Soviet Union. These flights produced a wealth of information about not only TALL KING but also other radars.

The third question was the sensitivity of the TALL KING receiver. Poteat proposed a system to electronically inject false airborne targets into radar systems. By receiving, modifying, and retransmitting the radar signals, a false target of a given size could be made to appear to fly along a defined course. Bud Wheelon, the Agency's deputy director for science and technology and the person responsible for OXCART, dubbed the spoofing system PALLADIUM. By analyzing air defense communications, the Agency operators would know when the false target was seen and thus determine how close OXCART could come before being detected. It also revealed the RCS that an aircraft needed in order to escape detection by various radars. PALLADIUM packages were used against a variety of Soviet radars, from ground locations, aircraft, and submarines. In one operation a submarine released a series of balloons carrying reflective spheres of known sizes in order to confirm the minimum RCS which the radar could detect [187].

More TALL KING information was obtained by mounting a miniature invasion of Cuba. In an armed strike, equipment and manuals were snatched from a radar installation. Although the Cuban military followed the invaders back out to sea, they escaped with their booty. In a bit of collateral damage, an uninvolved cargo ship was shot up by the Cubans. The ship turned out to belong to a company owned by John McCone, the Director of Central Intelligence, who upbraided Wheelon (Wheelon, A. D., speech to Roadrunners Internationale reunion, 2 Oct. 2003).

The results of all of the information gathering was bad news for OXCART. Poteat wrote,

> We had finished our special mission in support of the OXCART stealth program and gave our collected data, now called Quality ELINT, to the OSI analysts. The analysts then finished their vulnerability analysis job by concluding that the Oxcart would indeed be detected and tracked by the Soviets, which by then was no surprise to any of us. The OSI analysts put it to me differently, saying that we had just proved that the Earth was round and that, as soon as the Oxcart came over the horizon, the Soviet air defense radars would immediately see and track it. At the same time, we had also established realistic stealth radar cross section goals that, if met by the next generation of stealth aircraft, would allow the aircraft to fly with impunity right through the Soviet radar beams [187].

First Mission

Before the Agency's A-12s had even flown a mission, they became the focus of a controversy over whether they and the Air Force's upcoming SR-71s were redundant. After much discussion, a decision was made that the A-12s would be retired. However, in 1967 their capabilities were required in the Vietnam War. Since the Air Force's SR-71s were not yet qualified for operational use, three A-12s were deployed to Kadena, Japan, in an operation called BLACK SHIELD.

On 31 May 1967, when Agency pilot Mel Vojvodich flew the first mission to photograph North Vietnam, the vulnerability analysis by the ELO was borne out. In October 1995, he recounted the mission in a speech to the Roadrunners Internationale reunion:

> You know, this is the first stealth aircraft ever built and it's harder than hell to detect. But, the Japanese were sitting at the end of the runway when I took off and there were other people sitting there too, called in our takeoff time. I hit the tanker 100 miles downstream. They had a Russian trawler sitting right under our refueling track. They're saying "OK, he's leaving the tanker and he's heading for Hanoi or wherever he's going." So, when I made the turn there, they knew I was there.

My panel lit up with ECM gear and I knew they were gonna have a launch. They started launching SAMs ninety miles away. I could see 'em coming up. They were passing on the control down the different battalions to try to knock me out of the sky. The only way they could ever get an SR [*sic*] was to shoot way out in front of us.

Anyhow, many of them came up and you know, I'm not used to flying straight and level. I'm a fighter pilot and it's harder 'n hell for me to sit there, even at 85,000 feet and Mach 3.2 and some bastard's shootin' missiles at me.

I had the biggest urge to want to make a hard turn and get the hell out of there. But, I remember Slater [Col. Hugh 'Slip' Slater] said, "Now, just trust your ECM gear Mel, it's gonna work" and I'm sitting there saying "Yeah, I wonder why HE isn't here."

Anyway, it's kind of shaking to look at these missiles coming up through about 50,000 feet in con. They fired 24 at a time, coming up like this. I could see 'em coming up and they went underneath the belly of the aircraft and I had to look through my view periscope and I could see 'em go behind me and they were up over 90,000 feet and they came down and tried to track and by that time they had burned out and they'd detonate behind the aircraft and had never even gotten close [188].

When John Parangosky wrote "The OXCART Story" for the CIA's *Studies in Intelligence* (under the pseudonym Thomas P. McIninch), he stated that "There were no radar signals detected, indicating that the first mission had gone completely unnoticed by both Chinese and North Vietnamese," that no hostile action was taken against any of the first seven missions, and that the first SAM launch against an A-12 did not take place until 28 October 1967 [189].

Although there might have been a bit of hyperbole in Vovodich's story, he confirmed years later that there had been multiple missiles launched and added that he did not know why Parangosky denied it in the article (Vovodich, M., personal communication to author, Oct. 2001). Whatever the details, the first mission confirmed that advances in radars had surpassed the stealth technology of the early 1960s.

Chapter 17

CONCLUSION

The A-12 was the first of a family of Lockheed Mach 3 aircraft (Fig. 134). Out of a dozen different design studies, some never progressed beyond the conceptual stage, such as a carrier-based version of the A-12 (Fig. 135). However, three designs became actual aircraft.

F-12

Under the Air Force's Project KEDLOCK, Lockheed designed, built, and flew an interceptor variant of the A-12. It used the Hughes GAR-9 missile and the AN/ASG-18 pulse Doppler radar. The radar and missile had been initially tested in a YB-58 bailed to Hughes by the Air Force (Fig. 136). The B-58's pod was modified to carry and eject the missile. The design went through a number of iterations, including the AF-12 (Fig. 137) and the AF-112 (Fig. 138).

When Lockheed began to pursue the F-12 concept, Kelly Johnson sent A-12 test pilot Lou Schalk to recruit one of the Hughes pilots, Jim Eastham (Interview with J. D. Eastham, Rancho Palos Verdes, CA, 21 Dec. 2002) Eastham continued to fly the B-58 for Hughes while flying the A-12 for Lockheed. Eventually he made the first Mach 3 flight of the A-12 (Interview with N. E. Nelson, Palos Verdes Peninsula, CA, 2 March 2002), the first flight of the YF-12A (Fig. 139), and also flew the SR-71.

The Air Force awarded Lockheed a production contract for 92 airframes of a design designated the F-12B. However, the contract was soon cancelled, and the aircraft never made it beyond the mockup stage. (See Fig. 140.)

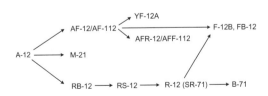

Fig. 134   A-12 family relationships. (Drawn by author from Whittenbury.)

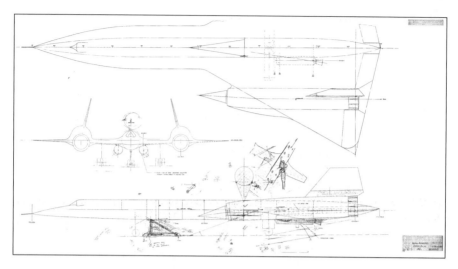

Fig. 135 A-12CB (carrier-based) concept, with catapult attachment and arresting gear, by Ed Baldwin. (Courtesy of the family of Edward P. Baldwin.)

Fig. 136 YB-58 with GAR-9 missile in pod. (Courtesy of James D. Eastham.)

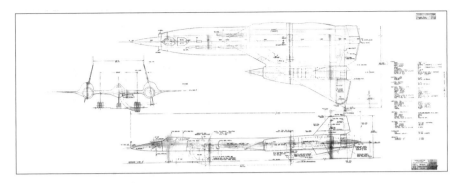

Fig. 137 Drawn in May 1961, the AF-12 was the first A-12 derivative for the interceptor role. (Courtesy of the family of Edward P. Baldwin.)

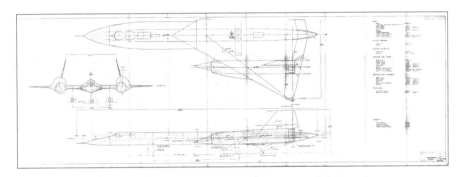

Fig. 138 Dating to March 1964, the AF-112D was another conceptual study for a Blackbird interceptor. (Courtesy of the family of Edward P. Baldwin.)

Fig. 139 First published photo of a Blackbird (YF-12A). (Courtesy of Roadrunners Internationale.)

Fig. 140   F-12B mockup, showing chines to nose. (Courtesy of Roadrunners Internationale.)

SR-71

The most successful was the Air Force's two-seat SR-71, originally designed under the designation R-12 (Figs. 141 and 142). It was the only Blackbird to make its first flight outside the Ranch, on 22 December 1964. Using a combination of cameras and synthetic aperture radar, it performed reconnaissance missions around the world for over 20 years. After its retirement, it served as a research testbed for NASA until its final flight in 1999.

"SR-71" – NOT A MISTAKE

For many years the story has been told that President Lyndon Johnson created the name "SR-71" during the public announcement by misreading his

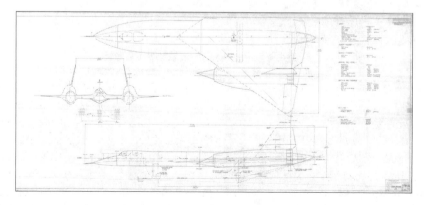

Fig. 141   R-12 general arrangement drawing, by Ed Baldwin. (Courtesy of the family of Edward P. Baldwin.)

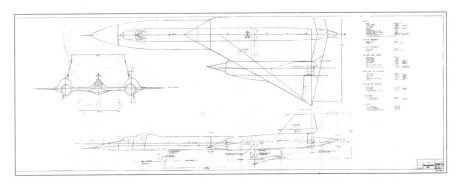

Fig. 142 TR-12 (SR-71B) trainer, by Ed Baldwin. (Courtesy of the family of Edward P. Baldwin.)

script, which supposedly said "RS-71." The story was published in the memoirs of Kelly Johnson and of Ben Rich and was told to this author by Norm Nelson, Jim Eastham, and numerous others.

Researcher John Wilson, of the Lyndon Johnson Presidential Library, has located various revisions of the script for the 24 July 1964 speech and has found that the story is not true. In fact the final script says "SR-71," and the President read it as written. The confusion came when a stenographer transcribed the speech and mistakenly wrote RS-71. The transcript was distributed to the press corps, who confused it with the script, and an urban legend was born [190].

Nevertheless, the origin of SR-71 remains a mystery. Early drafts refer to the SR-12, and in a recorded telephone call to the President two hours before the press conference, Secretary of Defense Robert McNamara refers to it as the R-12. Who marked up the transcript to change "SR-12" to "SR-71" is unknown, as is the person who directed the change, although there have been unsubstantiated claims that it was Air Force Chief of Staff General Curtis LeMay.

M-21

The final A-12 derivative to fly was the M-21 launch vehicle for the ramjet-powered D-21 drone (Fig. 143). Developed under Project TAGBOARD, the M-21 was a reinforced A-12 airframe with a second cockpit for a launch control officer (LCO). It performed three successful launches of the drone, but the fourth resulted in a midair collision and the loss of the aircraft and death of the LCO. The D-21 continued under Project SENIOR BOWL and was launched from a B-52 and boosted to altitude by a rocket. Eventually four missions were conducted over China. In two of those, the D-21 returned and parachuted its camera package into the Pacific Ocean, but in each case the package sank before being recovered.

Fig. 143 D-21 undergoing systems testing. (Courtesy of Lockheed Martin.)

A-12 Bomber Variants

Lockheed attempted to sell the Air Force on a bomber version of the Blackbird. In December 1960, while the A-12 was still over a year from its initial flight, the RB-12 was detailed in Technical Report SP-229, by W. W. Tjossem, K. Hoff, and W. M. Taylor. It would have used a rotary bomb dispenser in the fuselage. A much later variant was the B-71, which was derived from the R-12 (SR-71). It would have carried three free-fall bombs in each chine and would have used the same side-looking airborne radar (SLAR) as the SR-71 (Fig. 144).

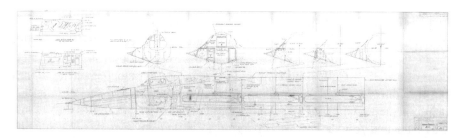

Fig. 144 This May 1966 inboard profile of the B-71 shows how a launcher would swing each bomb downward and into a slight nose-down attitude before release. (Courtesy of the family of Edward P. Baldwin.)

Convair WO 540

In November 1963, the same week that President Kennedy was assassinated, Convair began work on a proposal for a successor to the A-12. Bob Widmer assigned Randy Kent to be the project leader. Known simply as "Work Order #540," the concept was based on FISH. Powered by Marquardt ramjets burning hydrocarbon fuel, the parasite would have had variable-sweep wings. Work continued into July of 1964, at which point it was cancelled. Work continued on internal R&D funding and evolved into a Mach 6 design (Interview with R. Widmer and R. Kent, Fortworth, TX, 7 Nov. 2004).

Lockheed D-33

Another proposed successor to the A-12 was the Lockheed D-33. The project leader was Norman Nelson, whom Bissell had brought into Lockheed as the Agency liaison when production of the first A-12 encountered delays. In 1966, Nelson and James Reichert set a goal of a Mach 2.5 cruise at 100,000 ft with a range of 5,000 n miles. They wanted a small vehicle that could be launched from a variety of locations, including aircraft carriers and small runways. The resulting aircraft had delta wings with a planform similar to that of the Concorde, but was only 87 ft long and had a wing span of 34 ft. It had a takeoff gross weight of only 28,000 lb and a landing weight of 16,000 lb.

The aircraft was powered by a Pratt and Whitney Model 304 engine and for the first two minutes was also boosted by a P&W RL-10 rocket engine (Fig. 145). During acceleration and climb, it would burn 4,500 lb of liquid

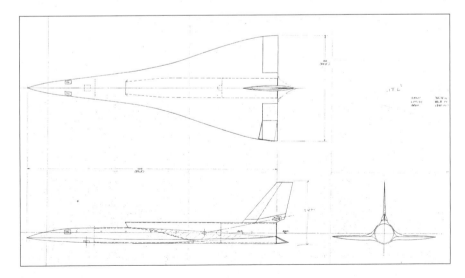

Fig. 145 D-33, showing Model 304 engine and RL-10 rocket. (Courtesy of Norman E. Nelson.)

hydrogen and liquid oxygen, taken from external tanks that would be dropped when the fuel and oxidizer were exhausted. In cruise it would burn liquid hydrogen and use external air taken in via an inlet on the top of the fuselage. An alternative design used a J58 burning liquid hydrogen.

The D-33 carried a single pilot, the Hycon camera used in the D-21 drone, and smaller versions of existing electronic countermeasures (ECM) gear, such as Big Blast, Blue Dog, and the Mad Moth system used on the A-12. Narmco provided information on radar-absorbent materials, and EG&G performed RCS measurements on models.

The proposal for the D-33 was not accepted by the CIA. Whether the designers would have been able to achieve the desired range is unknown.

Project ISINGLASS

An A-12 follow-on that received a great deal of attention in the mid-1960s was a system known as ISINGLASS. Concerned that the A-12 would be vulnerable to missiles if radars could be networked to give sufficient warning, in 1964 the CIA solicited proposals from McDonnell Aircraft. Using corporate funds, McDonnell designed a Mach 20 boost-glide aircraft, and Pratt and Whitney developed the XLR129 rocket engine. The 132,000-lb vehicle would have been launched from a B-52, although the practicality of that is uncertain [191].

By November 1965, the program had progressed to the point where the National Reconnaissance Office (NRO) had agreed in principle to fund further studies from its general R&D account [192]. Work continued into 1968, but eventually the project was cancelled.

B-2 Stealth Bomber

Although it was not a reconnaissance aircraft, the B-2 "Spirit" stealth bomber deserves mention because of its heritage from the early stealth work. In fact, Frank Rodgers was involved with its design. He apparently was still a believer in avoiding straight lines because years later he complained to Norm Taylor that he had tried unsuccessfully to have the many straight lines removed, especially the straight leading edge on the wings (Interview with N. H. Taylor, Topsfield, MA, 13 Oct. 2003).

Summary

Although the A-12 never met the original project goals of flying over the Soviet Union while being invisible to radar, it was nevertheless a success in a number of ways. At the tactical level, it provided valuable intelligence from Southeast Asia. It was the first member of a family of aircraft that provided strategic intelligence for over two decades and flight research data until the end of the century. And at the technological level, it broke new ground on airframe, engines, systems, and stealth technologies that continue to evolve.

The creation of the A-12, like the creation of the U-2 before it, was made possible by a number of factors that occur only rarely in the aerospace world. The most important was the existence of an urgent national need; intelligence essential to the nation's survival was needed, and this motivated individuals and institutions. A few visionary individuals in the government and industry with a willingness to take risks and the ability to think creatively accepted the challenge. There was a great deal of trust between the government and the contractors, which permitted the work to go forward with a minimum of oversight and bureaucratic delays. [Henry Combs estimates that for every customer engineer monitoring a contract, the contractor must devote one engineer full time to providing information to the monitor (Interview with H. Combs, Santa Clarita, CA, 19 April 2003).] The degree of trust meant that the failures and delays that are inevitable in a complex project would be tolerated and the project would go forward. The use of highly competent personnel meant that they could accept a great degree of responsibility for their work and could proceed without detailed guidance or second guessing. And finally, the extreme secrecy allowed the project to reach the operational stage before effective countermeasures could be developed [193].

"Be quick. Be quiet. Be on time."
—Kelly Johnson

Appendix A

SKUNK WORKS ENGINEERING STAFF

Although thousands of people were involved in developing, building, and testing the A-12 and its many systems, the core engineering staff of the Skunk Works consisted of approximately 75 people. Following is a list of most of those involved: Clarence L. "Kelly" Johnson; Verna Palm, secretary; Dick Boehme; Henry Combs; Don Nelson, aerodynamics; Dick Fuller, aerodynamics; Dick Cantrell, aerodynamics; Dave Campbell, thermodynamics; Ben Rich, thermodynamics; Edward Baldwin, structures; Sam Kelder, structures; Leroy English, structures; Ray Kirkham, structures; Herb Nystrom, structures; Bob "Flutter" Murphy, structures; Dan Zuck, structures; Ray McHenry, structures; Doug Wakefield, structures, cockpit; George Soto, canopy; Alvin Jensen, lofting; Ed Seitz, weight; Jack Koga, basic loads; Lorne Cass, basic loads; Merv Heal, weight; Tom Takesugi, air conditioning; Al Lehrer, stress; Bruce Galt, stress; Tom Pollack, stress; Bob Batista, stress; Rene Laurencot, stress; Gus Dishman, stress, landing gear; Herman Karlsson, landing gear; Cornelius "Corny" Gardner, landing gear; Ole Bendicson, landing gear; "Dag" D'Agostino, static testing; Bob Gavin, static testing; J. F. Campbell, materials; Vic Rummel, materials; Bob James, sheet metal design; Bob Wiele, sheet metal design, wings; Cliff Willoughby, sheet metal design; Roy Dow; Bill Bissell; Bob Charlton, illustrator; Frank Bullock, cockpit; David Robertson; Vern Bremberg, hydraulics; Leon Gavette, ground handling equipment; Sam Vose, ground handling equipment; Ed Martin, manager of functional design; Chris Fylling, functional design, wing edges; Vic Sorensen, functional design; Jack Painter, functional design; Don Bunce, functional design; "Rocky" Rockel, electrical design; Benson, electrical design; Doug Cone, air conditioning; Elmer Gath, propulsion; John Cadrobbi, propulsion; Davis, electrical design; Carl Allmon, lofting; Alvin Jensen, lofting; Herb Ermer; Willy Damwyck; L. D. MacDonald, electromagnetics; Mel George, chemistry; Edward Lovick, electromagnetics; Perry Reedy, chemistry; Ray Burton, chemistry; Michael Ash, electromagnetics; and James Herron, electromagnetics.

Appendix B

TIMELINE OF PROJECTS RAINBOW AND GUSTO

TABLE B   TIMELINE OF VARIOUS EVENTS AND DOCUMENTS RELATED TO PROJECTS AQUATONE, RAINBOW, AND GUSTO

Date	Event or document
1955/01/29	U-2 GA drawing, revision A
1955/07/12	Consideration of drone version of U-2
1955/11/09	CLJ visits Pentagon to pitch "L-182," including LH2 version.
1956/01	CLJ visits Putt et al. with SUNTAN proposal.
1956/01/18	Putt holds meeting to evaluate SUNTAN proposal.
1956/02/20	Pratt & Whitney selected
1956/02/20	Bill Sens proposed H_2 engine requirements.
1956/06/20	First operational flight of U-2; tracked, but altitude misread at 42 kft.
1956/06/21	Killian, Land, and Bissell met Goodpaster about expected yield.
1956/07/02	Eisenhower asked Goodpaster to ask Bissell whether tracked.
1956/07/02	Second and third U-2 operational flights
1956/07/03	Eisenhower told Goodpaster to have tracking reported.
1956/07/04	First U-2 flight over USSR; MiGs attempted to intercept
1956/07/05	Second U-2 flight over USSR
1956/07/05	Eisenhower says to suspend if evidence of tracking
1956/07/09	More U-2 missions
1956/07/10	MFR by Goodpaster; Eisenhower says proceed with operations until first report of tracking.
1956/07/10	Protest note by USSR; flights suspended.
1956/07/13	Bissell memo to Goodpaster; inactive for at least a week; talk when Eisenhower returns to Washington, D.C.
1956/07/17	Summary of results of initial overflights by H. I. Miller
1956/07/19	Dulles reported no operations are in progress.

(*Continued*)

TABLE B TIMELINE OF VARIOUS EVENTS AND DOCUMENTS RELATED TO PROJECTS
AQUATONE, RAINBOW, AND GUSTO (CONT)

Date	Event or document
1956/08/early	Bissell meets Land panel to urge radar detection reduction (same as 8/16 meeting?).
1956/08/06	Patent 3,127,608 on plasma stealth
1956/08/16–17	Meeting about reducing RCS: Johnson, Bissell, Purcell, Land, S. Miller, H. Miller
1956/08/17 or 18	First meeting of Frank Rodgers, Bob Naka, and Tom Bazemore with Edwin Land
1956/09/17	Eisenhower w/ Radford, Cabell, Bissell, Bridge; reviewed results of July operations
1956/10/1–6	T-33 w/ wires tested at Eglin AFB, FL
1956/10/03	Eisenhower meeting with Bissell et al. expressing discouragement at detection
1956/10/29	Bissell advises Purcell or Rodgers of Air Force program on reducing reflectivity by ionizing air around vehicle.
1956/12/14	Evaluating dirty birds at Indian Springs
1956/12/22	Sigint mission 4019 with System 5
1956/12/23	First LH2 B-57 flight in Project Bee
1957/03/18	Sigint mission 4020 [4030?] with System 5 to collect radar frequency data
1957/04/02	Article 341 crash; Bob Sieker killed
1957/05/02	Termination of DoD study group on implications of RAINBOW on national defense
1957/05/06	Bissell meeting with Eisenhower reporting progress with RAINBOW
1957/05/29	Navy, USAF, et al. briefing at Pentagon on AQUATONE (Geary) and RAINBOW
1957/middle	CLJ recommends cancellation of SUNTAN to James H. Douglas (Secretary of AF).
1957/07/01	John H. Collins, NACA Lewis, visited ADP. CLJ mentioned L/D for Mach 2 – 2.5 70,000-ft aircraft, which would be around for 10 years.
1957/07/21	First Dirty Bird flight along Black Sea coast; Pilot Cherbonneaux; Mission 3030
1957/07/31	Second Dirty Bird flight
1957/08/05	Dirty Bird flight over USSR space launch facility
1957/08/07	Discussion of withholding DMRs from some individuals
1957/08/23	Cabell, Bissell, and Twining meet Eisenhower to report on SOFT TOUCH and RAINBOW results

(Continued)

APPENDIX B

TABLE B TIMELINE OF VARIOUS EVENTS AND DOCUMENTS RELATED TO PROJECTS AQUATONE, RAINBOW, AND GUSTO (CONT)

Date	Event or document
1957/08/27	Discussion of withholding CMRs from some at NSA; concern that use of RAINBOW would be apparent
1957/09/06	Passive ECM Committee (FAR reported problems at Lockheed applying RAM)
1957/09/11	P&W Model 304 engine tests begin
1957/10/01	Paper on Soviet intercept capability; gives specific concerns and analysis of completeness of coverage
1957/10/02	Estimate of Soviet bloc high-altitude interception capabilities; analysis of tracking and intercept capabilities; area of likely detection increased
1957/10/03	Meeting in Cambridge; first discussion of new aircraft under Project RAINBOW; included presentation of "high aspect ratio flying wing with plastic empennage" being evaluated at Westinghouse
1957/10/24	Project Aircraft assignment; six allocated to RAINBOW
1957/10/24	Westinghouse authorized to construct 2400-ft antenna range on their property
1957/11/14	Visit to Headquarters, ARDC, by project director and deputy project Director; discussed AQUATONE in general and Thermos status, success, and future R&D; Gerald White assigned as liaison, to ensure project office and ARDC kept up to date on relevant work
1957/11/27	Paper reviewing status of RAINBOW Phase II; four techniques discussed; experiment organization not orderly
1957/12/02	Meeting on Project RAINBOW Status at ARDC. SAC had requested info on Thermos. Keeping [the Ranch] open.
Late 1957	CLJ sketches of B2 concept
1957/12/03	B2 GA drawing
1957/12/04	Meeting in Cambridge on specific tasks; see also tentative conclusions document; realized that internal components would have significant reflection
1957/12/09	All-metal airplane GA drawing
1958/01/02	Study of estimate of Soviet bloc high-altitude interception capabilities; Note changes from October and February 1957 estimates.
1958/01/08	Elliptical wing study, GA #2
1958/01/13	Status report Rainbow program 13 January 1958; mentions getting better measurements of "wires" in preparation for Phase II; also "sharks teeth"
1958/01/17	White House notes on special meeting; short operation; maybe don't use "covering" if range too short

(Continued)

TABLE B   TIMELINE OF VARIOUS EVENTS AND DOCUMENTS RELATED TO PROJECTS AQUATONE, RAINBOW, AND GUSTO (CONT)

Date	Event or document
1958/01/29	Rod Scott et al. of Perkin-Elmer visited project HQ; would be involved in GUSTO
1958/01/30	CLJ letter to Dick proposing work statement and cost estimate for GUSTO
1958/02/04	Trapezoidal wing study, GA #2
1958/02/05	GA-2 drawing
1958/02/11	Bissell approves CLJ proposal
1958/02/13	Goodpaster told Bissell to go ahead with "preliminary operations"
1958/03/01	Dirty Bird flight 6011 tracked; protest note sent 5 March; Eisenhower halts overflights 7 March.
	21 March aide-memoire contained detailed tracking information
1958/03/11	Comparison of measurements on RAM from Netherlands with WADC measurements of the same
1958/03/12	Fuselage contours drawn for GA #2
1958/03/??	#355 tracked by USSR; No Trapeze installed.
1958/03/31	Boom calculations, GA #2
	Batplane (Kirkham)
1958/04/03	Decision to continue testing at Indian Springs and not move to the Ranch
1958/04/09	Scimitar wing design, GA #3
1958/04/10	Memo on radar-absorbent [sic] material, Foote; need to brief key DoD personnel on use and threat from radar-absorbent materials
1958/04/10	Bissell visited [Land? SEI] in Boston
1958/04/10	Gibbs visited Owens-Corning. They've never done highly loaded Fiberglass structures.
1958/04/15	Gibbs visited Wright Field Materials Lab. Aircraft industry can't do Fiberglass structures; Status of specific materials; recommend Lockheed do GUSTO metal design and Goodyear or [...] build plastic parts. If passed static tests, then fly.
1958/04/21–23	"Basic Approach to Design of U-3"
1858/04/24	Goodpaster advised Twining and Dulles that Eisenhower says no recon over USSR.
1958/04/30	Report from Mr. C. L. Johnson on Project GUSTO, to Kiefer. CLJ called 4/28 to report on GUSTO II; probably won't meet requirements; proposed supersonic design

(*Continued*)

APPENDIX B

TABLE B TIMELINE OF VARIOUS EVENTS AND DOCUMENTS RELATED TO PROJECTS AQUATONE, RAINBOW, AND GUSTO (CONT)

Date	Event or document
1958/05/15	Estimates of required performance for GUSTO, Rodgers/Purcell (?); could probably come up with a GUSTO aircraft effective against S-Band
1958/05/15	Discussion of radar cross section of aircraft as function of range and altitude; edited draft of blip-scan study report?
1958/06/12	Air Council review of LH2 proposals by Lockheed, North American, Boeing, and Convair
1958/06/19	"Study of Configurations, U-3"
1958/06/26	"Interim Eng. Situation"
1958/06/26	"Evaluate Dry vs A.B. Versions of J-58"
1958/06/26	Archangel first concept drawing, Kelly Johnson
Undated	F-108 plan and side view sketches One side view shows missiles. Formerly available at http://www.wpafb.af.mil/museum/research/fighter/F108a-4.jpg
1958/06/30	"Cost Study – Project G-2" [also says "Project G"]
1958/07/02	Trip Report—25 June 1958 – 27 June 1958, Kiefer; at ADP, discussed supersonic and subsonic designs
1958/07/03	"General Arrangement: Four Ramjet Super Hustler," FW5810047, Convair
1958/07/early	Last P&W Model 304 engine runs
1958/07/03	A-1 GA drawing
1958/07/23	Presented Archangel I and Gusto G2A to Program Office
1958/08/12	DCI approved budget for GUSTO, CORONA, and [ ]
1958/08/13	Discussion of removing antiradar treatment from the last two U-2s that have it
1958/08/14	Discussions at Program Office; gave CLJ info on inflatable airplane and balloon combo
1958/08/15	Eisenhower approved one or two missions in the Far East.
1958/08/18	Study of launch alternatives (tow, balloon, rocket) On Archangel: Put on ram-jet tip tanks; what does staging do?
1958/08/22	Peterbilt GA drawing
1958/08/23	High-wing Ram Jet Kite GA drawing
1958/08/25	Got ramjet data from Marquardt and P&W; Added wing-tip ramjets to Archangel as per 1954 F104 proposal to USAF
1958/08/25	Mid-wing Ram Jet Kite GA drawing
1958/08/25	"Data from Marquardt" 150,000-ft ramjet

(Continued)

TABLE B   TIMELINE OF VARIOUS EVENTS AND DOCUMENTS RELATED TO PROJECTS
AQUATONE, RAINBOW, AND GUSTO (CONT)

Date	Event or document
1958/08/27	"A Study on Getting Archangel II to 100,000' Cruise Altitude by Going to M = 3.2 plus Ram-jets on Tip" "Revise Archangel to Higher Speed & Alt."
1958/08/27–28	Report outline, SP-100 (comparison of three ramjet-based designs)
1958/09/03	"Bill Sens – P&W J-58 Data"
1958/09/03	A-2 GA drawing
1958/09/04–08	Continuation of SP-100
1958/09/09	"What Factors Are Required To Make a 20,000# Airplane Work"
1958/09/09	"Balloon Data" Notes referring to George Schenk, with balloon diameter computations
1958/09/09	"Report on Archangel I & II," SP-101
1958/09/11	"Design Study: Archangel Aircraft, SP-101"
1958/09/17–24	CLJ in Washington
1958/09/20	"Evaluate Wt. of P&W Lite-Wt. Power Plant" [pages 2 & 3 missing; requested from David Lednicer 5/17/2001] "Evaluate a 10,000# 135,000' M = 3.0 Aircraft" "Try a Borane Job" (try to eliminate fuselage except for cockpit and equipment bay)
1958/09/21	"Try U-2 for Tow Job"
1958/09/22–23	Boston to review Archangel Project • Presented report on Navy inflatable design • Convair proposed Super Hustler • CLJ presented Gusto 2A • Archangel II presented and rejected because of penta-borane and cost Told to do a sanity check on FISH
1958/09/24	Decided to scale down Archangel II to 17,000 – 20,000 lbs and use JT-12A
1958/09/29	"Design of A-3" Basic concept—reduce radar C.S.
1958/10	SUNTAN project curtailed
1958/10	A-3 variant GA drawing
1958/10/03	"Weight Breakdown Given Project to Aim for"
1958/10/09	"Thrust Values Req'd for A-3"
1958/10/10	Unnamed A-3 GA drawing (Cherub #2 ?)
1958/10/13	Cherub #1 GA drawing
1958/10/14	Cherub variant GA drawing

(*Continued*)

APPENDIX B

TABLE B   TIMELINE OF VARIOUS EVENTS AND DOCUMENTS RELATED TO PROJECTS AQUATONE, RAINBOW, AND GUSTO (CONT)

Date	Event or document
1958/10/21	"Development Plan – A-3"
1958/10/22	"Data from Bill Sens"
1958/10/24	Nomograph of crippling strength of Titanium, by Batista
undated	"Perry Pratt – Ram Jet"
1958/10/27	"Perry Pratt – JT-12A"
	"Will have check soon"
	"Data to Ray M. – Oct. 9, 1958" [out of order in notebook]
1958/10/30	A-3 GA drawing
undated	RCS studies
1958/11/10	"Proposal for a Lightweight Reconnaissance Aircraft, SP-108"
1958/11/10	CLJ cost estimate for two G2A aircraft for flight evaluation of RCS; Will visit this week.
1958/11/15	Memo for Dr. James R. Killian, Land/Perkins/Purcell/Donovan/ Stever Recommend Super Hustler over Lockheed design
1958/11/21	Convair contract amendment #3; Continue [...] studies as well as lower-altitude conventional metal aircraft; Reports due 2/15/59
1958/11/26	"Call on Wed, Nov. 26, 58, D. B. on new proposal"
	Convair 1st choice, conditional on radar
	Frank Rogers [sic – Rodgers ?] coming down
1958/11/26	"Cost Estimate A-3 – Tunnel Tests & Radar tests"
1958/11/26	"Concept of A-4"
1958/12/02	Wing area calculation, A-4
1958/12/03	Arrow I GA Drawing
1958/12/03 – 1958/12/09	"Further Studies of A-4"
1958/12/05	A-5 GA drawing
1958/12/11	A-4 GA drawing
1958/12/12	B-58 Launched Vehicle GA drawing
1958/12/12	B-58 Launched Vehicle on B-58 GA drawing
1958/12	Arrow I and B-58 launched vehicle
1958/12/22	John S. D. Eisenhower, MCP: Hull, Conolly, Darden, Doolittle, Lovett, Cassidy, Killian, Gray, Goodpaster, J. Eisenhower, re: Eisenhower questioning continuation of overflights, 1 pg; mentions successor as higher performance
1958/12/24	Radar program [ ], Kiefer; $\frac{1}{8}$-scale model of "unsoftened wing"; also refers to "bag"

(*Continued*)

TABLE B TIMELINE OF VARIOUS EVENTS AND DOCUMENTS RELATED TO PROJECTS AQUATONE, RAINBOW, AND GUSTO (CONT)

Date	Event or document
1958/12/29	"Quote on 3-mo. Engineering on A-5"
1959/01	A-7 through A-9
1959/01/06	A-7-1 GA drawing
1959/01/07	A-7-2 GA drawing
1959/01/09	A-6-5 GA drawing
1959/01/12	Meeting at Convair
1959/01/15	A-7-3 GA drawing
1959/01/15	Bissell cover letter for report on investigations into redacted field; RAINBOW? More research needed than originally thought
1959/01/15	Comparison study of proposed follow-on vehicle [SH v. A-3], Director of Opns. Super Hustler better than A-3
1959/01/15	"Follow-On" operational considerations
1959/01/15	"Follow-On" evaluation criteria
1959/01/16	"Follow-On" range criteria, director of operations
1958/01/20	A-6-6 GA drawing
1959/01/29	Reconnaissance guidelines for GUSTO; Discussion of package configurations
1958/01/30	A-6-9 GA drawing
1959/02/04	Photographic configurations for GUSTO; Discussion of capabilities of EK
1959/02/04	Re: MFR: contract negotiations at CONVAIR, Bissell; Needs decision by Contracting Officer, DPD.
1959/02/12	John S. D. Eisenhower, MFR re: meeting with McElroy, Quarles, and Twining re: reconnaissance over USSR, 12 February 1959, 2 pp; Request for more overflights; U-2 successor coming along nicely—1–2 years; eight satellite flights in 1959
1959/02/12	Project R&D officer visited Eastman Kodak
1959/02/13	A. J. Goodpaster, MCP on 10 February 1959, Killian, Purcell, Land, Goodpaster, 13 February 1959, 3 pp; monitoring missile launches via sound, domestic demagogues, Corona (Land), high-altitude, high-performance recon aircraft (Purcell: saucer best shape, hard to track – blip-scan; Land: 700-lb payload, unseen until dropped payload), missile programs
1959/02/13	Supersonic in-flight refueling; summary of 1959/02/03 meeting with CLJ about refueling of A-7-2; feasible but not practical; Wing-tip refueling can't be done with A-7-3. [A-7-2 is low wing; -3 is high wing.]
1959/02/20	Bissell memo about EK visit; three camera manufacturers solicited informally for proposals

(*Continued*)

APPENDIX B

TABLE B   TIMELINE OF VARIOUS EVENTS AND DOCUMENTS RELATED TO PROJECTS AQUATONE, RAINBOW, AND GUSTO (CONT)

Date	Event or document
1959/02/27	DPD-DD/P staff meeting on procurement of Super Hustlers and ancillary materiel
1959/02	A-10
1959/02	SUNTAN cancelled
1959/03/05	Full-pressure altitude suit assemblies for GUSTO; Get David Clark Co. specs on X-15 suit and add GUSTO requirements to X-15 requirements as cover.
1959/03/06	Westinghouse contract BE-2022 completed; disposition of equipment
1959/03/9–10	Convair visit to DPD with statement of work
1959/03/14	Bissell disapproves contract
1959/03/11 ? 12	$40 k Contract with Firewell Corporation for pilot protective system
1959/03/12	A-11 GA drawing
1959/03/20	A-11A GA drawing
1959/03/25	Reconnaissance photography; refers to study by EK [Kiefer?] sent to Land Panel
1959/04/28	A-11 wing-tip fins sketch
1959/05/15	Summary—GUSTO Program as of 15 May 1959, Kiefer; no testing of Lockheed configuration; details on Convair configuration and redesign; Navy funding J-58 through next year.
1959/05/25	"Follow-On" operational considerations, director of operations
1959/05/25	"Follow-On" evaluation criteria
1959/05/26	Comparison study of proposed follow-on vehicle, chief, Operations Branch, DPD; Super Hustler vs A-11 point-by-point comparison
1959/06/05	[Selection of follow-on vehicle], Burke; A-11 chosen over Super Hustler
1959/06/09	Detection of proposed vehicle, USN Intel Officer; concludes that vehicle will be detected and tracked
1959/06/10	CIA/HQ USAF relationships re GUSTO, Burke
1959/06/26	Goodpaster (?) note that he called Bissell to reevaluate last six months' work especially with respect to "lowest (?) detection"
1959/07/01	Convair redirected to KINGFISH configuration
1959/07	KINGFISH
1959/07	Original A-12 GA drawing
1959/07	A-12 initial configuration
1959/07/03	Bissell visited Johnson. Said they'd extend program and accept lower altitude if aircraft adapted to reduce RCS. CLJ proposed A-12.
1959/07/20	Dear Doc. Summary of optical systems.

(Continued)

TABLE B TIMELINE OF VARIOUS EVENTS AND DOCUMENTS RELATED TO PROJECTS
AQUATONE, RAINBOW, AND GUSTO (CONT)

Date	Event or document
1959/07/20	A. J. Goodpaster, "MCP: A. Dulles, Cabell, Bissell, McElroy, Kistiakowsky, Killian Goodpaster, 20 July 1959, 3 pp; review of U-2 successor; Super Hustler feasible. Lockheed chosen because of air-launch difficulty of Hustler. Performance specs; first flight January 1961; Dulles suggested bomber version; $6.5M spent on Convair. $100k spent on Lockheed proposal.
1959/07/31	Proposed name change from GUSTO [to OXCART], security officer, DPD
1959/08/19	Project GUSTO; conflict between camera installation and RCS shielding
1959/08/20	LADP and Convair submitted final proposals.
1959/08/28	Saw program office director: LADP had won.
1959/08/29	Go-ahead for $4.5 million for 1959/09/01–1960/01/01
1959/08/31	Started mock-ups and reorganization
1959/08/31	Memorandum for AFCIG-5, acting chief, DPD
1959/09/01	Start date
1959/09/10	Cancellation notice of Project GUSTO, chief, cover section, DPD
1959/11/02	Meeting between CLJ and Bissell
1959/11/06	"Dear Dick" letter to Bissell; alludes to 2 Nov. meeting
1959/11/09	Full-scale model complete
1960/02/08	A-12 GA drawing
1960/02/08	A. J. Goodpaster, MFR: Eisenhower and Board of Consultants, 8 February 1960, 2 pp; overflights in general; mention of new recon aircraft
1960/06/02	Eisenhower considering cancelling or reducing priority
1960/06/03	"Notes on OXCART," 3 June 1960, 2 pp; withdrawal sheet lists WH/USAF; First flight May 1961; operational spring 1962
1960/06/11	Note, 11 June 1960, 1 pg; "President thinks gen. OK to go ahead – Little chance of peacetime use."
1960/07/19	Wrote proposal for AF-12
1960/08/17	Announced that will be late and over cost
1960/09/14	Start design of bomber version
1960/11/29	Memo for Gene [Kiefer], Rodgers? Work on RCS
1960/12/15	RB-12 proposal
1961/05/23	AF-12 GA drawing
1962/03/05	Project GUSTO files; found in Bissell's safe

(*Continued*)

TABLE B TIMELINE OF VARIOUS EVENTS AND DOCUMENTS RELATED TO PROJECTS AQUATONE, RAINBOW, AND GUSTO (CONT)

Date	Event or document
1962/12/14	R-12 GA drawing
1963/01/09	Approval of final payment to Convair
1963/01/18	Final payment made to Convair
1964/02/18	TR-12 GA drawing
1964/03/05	AF-112D GA drawing
1965/02/24	A-12CB GA drawing
1965/03/19	B-71 GA drawing

Appendix C

SUPPORTING MATERIALS

GUSTO DOCUMENTS ARCHIVE

The GUSTO Documents Archive is a compilation of primary source documents pertaining to Projects RAINBOW and GUSTO, with a small number of documents pertaining to Projects AQUATONE and OXCART and related projects.

To download these supporting materials, please go to http://www.aiaa.org/publications/supportmaterials. Select your title, follow the instructions, provided and enter the following password: **rainbow**.

Many of the topics introduced in this book are discussed in more detail in other AIAA publications. A complete listing of titles in the Library of Flight, as well as other AIAA publications, is available at http://www.aiaa.org.

AIAA is committed to devoting resources to the education of both practicing and future aerospace professionals. In 1996, the AIAA Foundation was founded. Its programs enhance scientific literacy and advance the arts and sciences of aerospace. For more information, please visit www.aiaafoundation.org.

REFERENCES

[1] Hall, R. C., "Postwar Strategic Reconnaissance and the Genesis of Corona," *Eye in the Sky: The Story of the Corona Spy Satellites*, edited by Dwayne A. Day, John M. Logsdon, and Brian Latell, Smithsonian Inst., Washington, DC, 1998, Chap. 4, pp. 87–111.
[2] Pocock, C., *50 Years of the U-2*, Schiffer Publishing, Ltd., Atglen, PA, pp. 9–15.
[3] Land, E. H., "A Unique Opportunity for Comprehensive Intelligence," TS-115018, Office of Defense Mobilization, Washington, DC, 5 Nov. 1954.
[4] Dulles, A. W., "Reconnaissance," TS-103085, CIA, Washington, DC, date illegible.
[5] "Project Outline," TS-103219, CIA, Washington, DC, 7 Jan. 1955.
[6] Bissell, R. M., Jr., "Project OARFISH (A Sub-project of AQUATONE)," TS-103262, CIA, Washington, DC, 25 Feb. 1955.
[7] "Preliminary Analysis of Flight No. 1, System One – S-Band, 20 June 1956," SAPC-7333, CIA, Washington, DC, 26 June 1956.
[8] Goodpaster, A. J., "Memorandum for the Record," Office of the President, Washington, DC, July 3, 1956.
[9] Goodpaster, A. J., "Memorandum for the Record," Office of the President, Washington, DC, July 5, 1956.
[10] Goodpaster, A. J., "Memorandum for the Record," Office of the President, Washington, DC, July 6, 1956.
[11] Goodpaster, A. J., "Memorandum for the Record," Office of the President, Washington, DC, July 10, 1956.
[12] Miller, H. I., "Suggestions re the Intelligence Value of AQUATONE," TS-158354, CIA, Washington, DC, 17 July 1956.
[13] Goodpaster, A. J., "Memorandum for Record," Office of the President, Washington, DC, July 19, 1956.
[14] Johnson, C. L., "U-2 Log," Lockheed ADP, Burbank, CA, 16–17 Aug. 1956.
[15] Salisbury, W. W., "Absorbent Body for Electromagnetic Waves," U.S. Patent 2,599,944, filed 11 May 1943.
[16] "Horten Ho229," Wikipedia, http://en.wikipedia/wiki/Horten_Ho_229.
[17] Crispin, J. W., Jr., Goodrich, R. F., and Siegel, K. M., "A Theoretical Method for the Calculation of Radar Cross Sections of Aircraft and Missiles," Report 2591-1-H, University of Michigan Radiation Laboratory, July 1959.
[18] Gibbs, J. A., "Radar Maintenance Technicians for Rainbow Project," SAPC memo (number illegible), CIA, Washington, DC, 26 Oct. 1956.
[19] R&D Officer, "Report on R&D Tests," TS-158816, CIA, Washington, DC, 23 Jan. 1957.
[20] "RAINBOW Program, Phase II," TS-16978/R, CIA, Washington, DC, 27 Nov. 1957.
[21] Bissell, R. M., Jr., "Memorandum for the Record: Concurrence in Letter Contract No. BE-2022 with Westinghouse Electric Corporation, Baltimore, MD. Project RAINBOW," SAPC-16622, CIA, Washington, DC, 14 June 1957.

[22] Contracting Officer, "Letter Contract No. BE-2022, Amendment No. 1," SAPC-16892, CIA, Washington, DC, 27 June 1957.
[23] Contracting Officer, "Contract Approval," SAPC-20390, CIA, Washington, DC, 24 Oct. 1957.
[24] Contracting Officer, "Contract Approval," SAPC-22847, CIA, Washington, DC, 27 Dec. 1957.
[25] Bissell, R. M., Jr., "Concurrence in Definitive Contract No. BE-2022 with Westinghouse Electric Corporation, Project [redacted] (RAINBOW)," SAPC-23332, CIA, Washington, DC, 15 Jan. 1958.
[26] Admin. Contracting Officer, memo to sales engineer, SAPC-24266, CIA, Washington, DC, 11 Feb. 1958.
[27] Contracting Officer, "Disposition of Property under Contract No. BE-2022 with Westinghouse Electric Corp.," DPD-1455-59, CIA, Washington, DC, 6 March 1959.
[28] "Experiments, Measurements and Comments on Reducing K_a Band Radar Echoes from an [redacted] Aircraft Model," CIA, Washington, DC, undated.
[29] "AQUATONE Meeting, 9:30 AM, Monday, 6 May 1957, Briefing Notes for DCI," TS-166427/A, CIA, Washington, DC, 3 May 1957.
[30] Reber, J. Q., "Handling of DMR in the Face of RAINBOW," SAPC-16957-R, CIA, Washington, DC, 7 Aug. 1957.
[31] Gibbs, J. A., "Dissemination of CMR to Additional People at NSA," SAPC-16970/R, CIA, Washington, DC, 5 Sept. 1957.
[32] Goodpaster, A. J., "Memorandum for Record," Office of the President, Washington, DC, 23 Aug. 1957.
[33] Price, A., *The History of U.S. Electronic Warfare, Volume II*, The Association of Old Crows, Alexandria, Virginia, 1989, pg. 200.
[34] Appold, N. C., "Minutes of the First Meeting of the Joint USAF-USN-Lincoln Laboratory Committee on Passive ECM," C7-115854, Air Research and Development Command, 6 Sept. 1957.
[35] Assistant Director, Scientific Intelligence, "Estimate of Soviet Bloc High Altitude Interception Capabilities," TS-164825, CIA, Washington, DC, 2 Jan. 1958.
[36] Gibbs, J. A., "Visit to Headquarters, ARDC, by Project Director and Deputy Project Director," SAPC-16979/R, CIA, Washington, DC, 26 Nov. 1957.
[37] Bissell, R. M., Jr., "Proposed Advanced Reconnaissance System," TS-164671, CIA, Washington, DC, 19 Nov. 1957.
[38] "RAINBOW Program, Phase II," TS-16978/R, CIA, Washington, DC, 27 Nov. 1957.
[39] Gibbs, J. A., "Trip to Hqs, ARDC," 2 December 1957, SAPC-16980/R, CIA, Washington, DC, 2 Dec. 1957.
[40] Bissell, R. M., Jr., "RAINBOW Phase II Tentative Conclusions," TS-158825/R, CIA, Washington, DC, 10 Dec. 1957.
[41] "Additional Tasks Developed at 4 December Meeting," TS-158824/R, CIA, Washington, DC, 6 Dec. 1957.
[42] Gibbs, J. A., "Clearances for [redacted] Personnel," SAPC-16982/R, CIA, Washington, DC, 19 Dec. 1957.
[43] "Study of Soviet Air Defense Capability Against High Flying Aircraft," TS-164856, CIA, Washington, DC, 24 Jan. 1958.
[44] Director of Operations, "Study of Soviet Air Defense Capability Against High Flying Aircraft," TS-164856, CIA, Washington, DC, 24 Jan. 1958.
[45] Goodpaster, A. J., "Memorandum of Conference with the President, January 16, 1958," Office of the President, Washington, DC, 17 Jan. 1956.
[46] Project security officer, "Mail Box," GUS-0014, CIA, Washington, DC, 24 Feb. 1958.
[47] Johnson, C. L., GUS-0006, Lockheed ADP, Burbank, CA, 30 Jan. 1958.

[48] "Status Report Rainbow Program," CIA, Washington, DC, 13 Jan. 1958.
[49] Kiefer, G. P., "Latest Lockheed Schedule on Delivery of THERMOS Aircraft," SAPC-169??/R, CIA, Washington, DC, 20 Feb. 1958.
[50] Director of Operations, "Tracking of 355," SAPC-25019-R, CIA, Washington, DC, 14 March 1958.
[51] Kiefer, G. P., "[redacted] Testing at [redacted]," SAPC-16999/R, CIA, Washington, DC, 3 April 1958.
[52] Johnson, C. L., "Reduction of Radar Cross Section of Large High Altitude Aircraft," Proceedings of the 1975 Radar Camouflage Symposium, Oct. 1975, Wright Patterson Air Force Base, OH.
[53] Johnson, C. L., "Proposal for a Lightweight Reconnaissance Aircraft," SP-108, Lockheed ADP, Burbank, CA, 10 Nov. 1958.
[54] Gibbs, J. A., "Visit to Owens-Corning by Col. Gibbs, 10 April 1958," GUS-0020, CIA, Washington, DC, 11 April 1958.
[55] Gibbs, J. A., "Visit to Materials Lab, Wright Field," GUS-0021, CIA, Washington, DC, 16 April 1958.
[56] Lockheed executive vice president, "Cost and Schedule Proposal for a Special High Altitude Aircraft," CIA, Washington, DC, 29 Nov. 1955.
[57] Kiefer, E. P., "Report from Mr. C. L. Johnson on Project GUSTO," GUS-0013, CIA, Washington, DC, 30 April 1958.
[58] Gibbs, J. A., "Concept of Operations," DPD-0416, CIA, Washington, DC, 9 May 1958.
[59] Bissell, R. M., Jr., "Monetary Discussion, Programs Assigned to SAD/PD," DPS-2610, CIA, Washington, DC, 19 July 1958.
[60] "Staff Meeting Minutes – 22 July 1958," COR-0084, CIA, Washington, DC, 22 July 1958.
[61] "Discussion of Radar Cross Section of Aircraft as Function of Range & Altitude," GUS-0037, CIA, Washington, DC, 15 May 1958.
[62] "Estimates of Required Performance for GUSTO," GUS-0036, CIA, Washington, DC, 15 May 1958.
[63] Bissell, R. M., Jr., *Reflections of a Cold Warrior: From Yalta to the Bay of Pigs*, Yale University Press, 1996, pg. 132.
[64] *Super Hustler: A New Approach to the Manned Strategic Bombing-Reconnaissance Problem*, FZM-1200-20, Convair, Fort Worth, TX, 26 May 1958.
[65] "Letter Contract No. [redacted]," document number [redacted]-0002, CIA, Washington, DC, 22 June 1958.
[66] Contracting Officer, "Change of Project Funds Obligated Under Contract No. SS-100, Convair, San Diego, California, Project CHAMPION," DPD-2827-59, CIA, Washington, DC, 30 April 1959.
[67] Contracting Officer, memo to Convair authorizing overtime, document number [redacted]-0057, CIA, Washington, DC, 27 Aug. 1958.
[68] Kiefer, E. P., "Trip report – 25 June 1958 - 27 June 1958," GUS-0043, CIA, Washington, DC, 2 July 1958.
[69] "Staff Meeting Minutes – 5 August 1958," CHAL-0280, CIA, Washington, DC, 6 Aug. 1958.
[70] Bissell, R. M., Jr., "Financing of Special Projects - Fiscal Year 1959," DPS-3074, CIA, Washington, DC, 12 Aug. 1958.
[71] Bissell, R. M., Jr., "Identification of Special Projects," TS-155106, CIA, Washington, DC, 13 Aug. 1958.
[72] Johnson, C. L., "Design Study: Archangel Aircraft," SP-101, Lockheed ADP, Burbank, CA, 11 Sept. 1958.
[73] "A-12 Log," entry for 17–24 Sept. 1958, Lockheed ADP, Burbank, CA.

[74] Johnson, C. L., *History of the OXCART Program*, SP-1362, Lockheed ADP, Burbank, CA, 1 July 1968, pp. 2–3.
[75] "Staff Meeting Minutes – 25 September 1958," COR-0149, CIA, Washington, DC, 26 Sept. 1958.
[76] Johnson, C. L., "A-12 Log," entry for 24 Sept. 1958, Lockheed ADP, Burbank, CA.
[77] Johnson, C. L., "Proposal for a Lightweight Reconnaissance Aircraft," SP-108, Lockheed ADP, Burbank, CA, 10 Nov. 1958.
[78] Kiefer, E. P., "Advisory Panel Meeting," GUS-0060, CIA, Washington, DC, 22 October 1958.
[79] Land, E. H., et al., "Memorandum for Dr. James R. Killian," GUS-0070, CIA, Washington, DC, 15 Nov. 1958.
[80] White, L. K., Bissell, R. M., Jr., and Macy, R. M., "Memo of Understanding: Funding of Proj GUSTO in FY-1959-FY-1960," GUS-0073, CIA, Washington, DC, 16 Dec. 1958.
[81] Eisenhower, John, S. D., "Memorandum of Conference with the President, Dec. 16, 1958 – 9:00 AM," Office of the President, Washington, DC, 22 Dec. 1958.
[82] Director of Operations, "Comparison Study of Proposed Follow-On Vehicle," GUS-0086, CIA, Washington, DC, 15 Jan. 1959.
[83] Report on FISH and KINGFISH, Convair, Fort Worth, TX, 23 July 1959.
[84] Contracting Officer, "Meeting on Support of Project GUSTO," GUS-0146, CIA, Washington, DC, 27 Feb. 1959.
[85] Bissell, R. M., Jr., "Project GUSTO, re:Memorandum for DD/P (GUS-0156)," GUS-0163, CIA, Washington, DC, 5 March 1959.
[86] "Propulsion Review Meeting," CIA, Washington, DC, 20 March 1959.
[87] Whittenbury, J. R., *From Archangel to OXCART: Design Evolution of the Lockheed A-12, First of the Blackbirds*, Lockheed Martin Aeronautics Company, Palmdale, CA, 1999.
[88] "Supersonic In-Flight Refueling," GUS-0109, CIA, Washington, DC, 13 Feb. 1959.
[89] Goodpaster, A. J., "Memorandum of Conference with the President, February 10, 1959," Office of the President, Washington, DC, 13 Feb. 1959.
[90] Eisenhower, J. S. D., "Memorandum for Record, Office of the President, Washington, DC, 12 Feb. 1959.
[91] Burke, W., "GUSTO Feasibility Study (Office of Security Support)," DPD-6621-59, CIA, Washington, DC, 16 Nov. 1959.
[92] Kiefer, E. P., "Radar Program," GUS-0076, CIA, Washington, DC, 24 Dec. 1958.
[93] "Staff Meeting Minutes – 20 January 1959," CHAL-0536, CIA, Washington, DC, 22 Jan. 1959.
[94] Bissell, R. M., Jr., "Concurrence in Letter Contract No. HL-4646 [ ] with The General Dynamics Corporation (CONVAIR Division), Fort Worth, Texas, Project GUSTO," GUS-0094, CIA, Washington, DC, 21 Jan. 1959.
[95] "Task Change Record No. 1," GUS-0127, CIA, Washington, DC, 9 Feb. 1959.
[96] Contracting Officer, "Weekly Contract Report," DPD-1114-59, CIA, Washington, DC, 27 Feb. 1959.
[97] Contracting Officer, "Weekly Contract Report," DPD-1501-59, CIA, Washington, DC, 6 March 1959.
[98] "Task Change Record No. 2, Electronics Model Tests and Resultant Models Modifications," CIA, Washington, DC, 27 Feb. 1959.
[99] "Task Change Record No. 3," CIA, Washington, DC, 30 March 1959.
[100] Contracting Officer, "Contract No. HL-4646, Convair, Fort Worth, Statement of Work, Project GUSTO," GUS-0170, CIA, Washington, DC, 10 March 1959.

[101] Chief, R&D Branch, DPD, "GUSTO Facilities," GUS-0176, CIA, Washington, DC, 12 March 1959.
[102] Bissell, R. M., Jr., "Contract No. HL-4646, Approval for Convair to Contract for Architectural Drawings for New Building for Project GUSTO," GUS-0171, CIA, Washington, DC, 11 March 1959.
[103] Bissell, R. M., Jr., "Contract No. HL-4646, Approval for Convair to Contract for Architectural Drawings for New Building for Project GUSTO," GUS-0182, CIA, Washington, DC, 14 March 1959.
[104] Project Engineer, "Trip Report Convair," DPD-2313-59, CIA, Washington, DC, 6 April 1959.
[105] Contracting Officer, "Concurrence in Definitive Contract No. HL-4646 with General Dynamics Corporation, CONVAIR Division, Fort Worth, Texas, Project GUSTO," GUS-0211, CIA, Washington, DC, 9 April 1956.
[106] "Task Change Record No. 4," CIA, Washington, DC, 20 April 1959.
[107] Telex to Director, "Task Change Record No. 5," IN-09257, CIA, Washington, DC, 6 May 1959.
[108] "TCR No. 9," GUS-0298, CIA, Washington, DC, 9 June 1959.
[109] Telex to Director, "Task Change Record No. 7," IN-09218, CIA, Washington, DC, 7 May 1959.
[110] Telex to Director, "Task Change Record No. 6," IN-09219, CIA, Washington, DC, 7 May 1959.
[111] "T.C.R. No. 8," GUS-0283, CIA, Washington, DC, 26 May 1959.
[112] Contracting Officer, "Weekly Contract Report No. 10," DPD-2834-59, CIA, Washington, DC, 1 May 1959.
[113] Burke, W., "Comparison and Evaluation of Two Weapons Systems Designed to Meet CHALICE OXCART Operational Requirements," GUS-085, CIA, Washington, DC, 5 June 1959.
[114] Johnson, C. L., "A-12 Log," Dec. 1958–July 1959 entry, Lockheed ADP, Burbank, CA.
[115] "Staff Meeting Minutes – 16 June 1959," COR-0509, CIA, Washington, DC, 18 June 1959.
[116] "Staff Meeting Minutes – 23 June 1959," COR-0526, CIA, Washington, DC, 26 June 1959.
[117] "Task Change Record No. 11," CIA, Washington, DC, 1 July 1959.
[118] "A-12 Log," entry for Dec. 1958–July 1959, Lockheed ADP, Burbank, CA.
[119] "Staff Meeting Minutes – 7 July 1959," COR-0547, CIA, Washington, DC, 13 July 1959.
[120] "Staff Meeting Minutes – 16 July 1959," COR-0552, CIA, Washington, DC, 16 July 1959.
[121] Telex to Director, "Message for Eugene Kiefer," IN-03269, CIA, Washington, DC, 16 July 1959.
[122] Telex from Director, untitled, OUT-98126, CIA, Washington, DC, 17 July 1959.
[123] Telex to Director, "Task Change Record Nbr. 12," IN-03368, CIA, Washington, DC, 17 July 1959.
[124] Telex to Director, "Message for Eugene Kiefer," IN-04177, CIA, Washington, DC, 29 July 1959.
[125] Telex to Director, "Message for Gene Kiefer," IN-04519, CIA, Washington, DC, 4 Aug. 1959.
[126] "TCR No. 13," CIA, Washington, DC, 17 Aug. 1959.
[127] Telex to Director, untitled, IN-0458?, CIA, Washington, DC, 5 Aug. 1959.

[128] Telex to Director, untitled, IN-04956, CIA, Washington, DC, 11 Aug. 1959.
[129] Rich, B. R., "F-12 Series Aircraft Aerodynamic and Thermodynamic Design in Retrospect," *Journal of Aircraft*, Vol. 11, No. 7, July 1974, pp. 401–406.
[130] Kiefer, E. P., "Notes for Mr. Bissell: Minority Dissent on GUSTO," GUS-0347, CIA, Washington, DC, 13 July 1959.
[131] Kiefer, E. P., "Notes for Mr. Bissell re GUSTO additional factors bearing on decisions reached at 8 July meeting," GUS-0346, CIA, Washington, DC, 13 July 1959.
[132] Bissell, R. M., Jr., "Status of Project GUSTO," GUS-0348, CIA, Washington, DC, 12 July 1959, 3 pp.
[133] "GUSTO Summary," GUS-0356, CIA, Washington, DC, 18 July 1959.
[134] Goodpaster, Andrew J., "Memorandum of Conference with the President, 20 July 1959," Office of the President, Washington, DC, 20 July 1959.
[135] "A-12 Log," entry for 28 Aug. 1959, Lockheed ADP, Burbank, CA.
[136] "Task Change Record No. 14," CIA, Washington, DC, 10 Sept. 1959.
[137] "Task Change Record No. 15," OXC-0109, CIA, Washington, DC, 12 Nov. 1959.
[138] Telex to Director, IN-07162, CIA, Washington, DC, 11 Sept. 1959.
[139] Telex to Director, "Outline of Proposed Materials Program and Discussion of Materials on Hand or in Transit," CIA, Washington, DC, 22 Sept. 1959.
[140] "Task Change Record No. 16," CIA, Washington, DC, 17 Nov. 1959.
[141] Deputy Chief, Development Branch, "Termination of [Convair] (GUSTO / OXCART) Feasibility Studies," OXC-0272, CIA, Washington, DC, 2 Feb. 1960.
[142] Chief of Contract Administration, Convair, "Final Patent Report," DPD-2848-60, CIA, Washington, DC, 25 March 1960.
[143] Burke, W., "Disposition of Materials and Residual under Contracts Nos. [ ] and HL-4646, [ ] and CONVAIR Corporations, Project GUSTO," GUS-0401, CIA, Washington, DC, 19 April 1960.
[144] Bissell, R. M., Jr., "Memorandum for Chief, Contracts Branch, DP Division, Disposition of Materials and Residual under Contracts Nos. [ ] and HL-4646, [ ] and CONVAIR Corporations, Project GUSTO," GUS-0403, CIA, Washington, DC, 22 April 1960.
[145] Dan, Memo to John, GUS-0404-60, CIA, Washington, DC, undated.
[146] Technical Analysis Staff, "Memorandum for Chief of Contracts Branch, Completion of Contract No. HL-4646-CONVAIR," DPD-2722-60, CIA, Washington, DC, 6 April 1960.
[147] Chief of Contracts Branch, DPD, "Review of Technical Data to be retained by CONVAIR, Project GUSTO," GUS-0405, CIA, Washington, DC, 3 June 1960.
[148] Chief, Technical Analysis Staff, "Review of Technical Data to be Retained by CONVAIR, Project GUSTO," GUS-0407, CIA, Washington, DC, 8 June 1960.
[149] Contract Negotiator, "Trip Report 27–28 June 1960, CONVAIR," DPD-5298-60, CIA, Washington, DC, 6 July 1960.
[150] President – GD/Fort Worth, "Contractor's Release," Convair, Fort Worth, TX, 19 Jan. 1962.
[151] President – GD/Fort Worth, "Contractor's Assignment of Refunds, Rebates, and Credits," Convair, Fort Worth, TX, 19 Jan. 1962.
[152] Contracting Officer, OSA, "Final Payment, Contract No. HL-4646, Convair, Fort Worth, Texas," OSA-0172-63, CIA, Washington, DC, 9 Jan. 1963.
[153] Chief, Budget and Finance Branch, OSA-DD/R, "Contract HL-4646, General Dynamics Corporation, Convair Division," OSA-0367-63, CIA, Washington, DC, 18 Jan. 1963.
[154] Security Officer, "Proposed Name Change from GUSTO [to OXCART]," DPD-5219-59, CIA, Washington, DC, 31 July 1959.
[155] Burke, W., "Cancellation Notice of Project GUSTO," GUS-0384, CIA, Washington, DC, 31 Aug. 1959.

REFERENCES

[156] Chief, Cover Section, "Cancellation Notice of Project GUSTO," GUS-0385, CIA, Washington, DC, 10 Sept. 1959.
[157] Burke, W., "OXCART Program Management," OXC-0147, CIA, Washington, DC, 3 Dec. 1959.
[158] Johnson, C. L., "Progress Report #1," OXC-0029, CIA, Washington, DC, 17 Sept. 1959.
[159] Telex to Director, untitled, IN-09523, CIA, Washington, DC, 16 Oct. 1959.
[160] MacDonald, L. D., "Recommended Improvements to Test Facilities," Lockheed ADP, Burbank, CA, 19 Oct. 1959.
[161] Telex to Director, untitled, IN-08527, CIA, Washington, DC, 1 Oct. 1959.
[162] "OXCART 2 November Meeting Agenda," CIA, Washington, DC, 2 Nov. 1959.
[163] Johnson, C. L., "Dear Dick," letter to Richard Bissell, OXC-0095-59, Lockheed ADP, Burbank, CA, 6 Nov. 1959.
[164] Rodgers, F. A., memo for Gene Kiefer, CIA, Washington, DC, 29 Nov. 1959.
[165] A-12 Log, entry for 16 Nov. 1959, Lockheed ADP, Burbank, CA.
[166] Parangosky, John, "OXCART - Pratt & Whitney Proposal for Development of Exhaust Ionizing Equipment for the J-58 Engine," OXC-0342, CIA, Washington, DC, 25 Feb. 1960.
[167] Johnson, C. L., "Progress Report #3," OXC-0145, Lockheed ADP, Burbank, CA, 30 Nov. 1959.
[168] Letter from Dan to Herb, DPD-8221-59, CIA, Washington, DC, 2 Dec. 1959.
[169] Johnson, C. L., "Progress Report # 4," OXC-0208-60, Lockheed ADP, Burbank, CA, 29 Dec. 1959.
[170] Dulles, A. W., "Memorandum for Director of the Budget, Project [redacted]," attachment to OXC-0323-60, CIA, Washington, DC, 30 Jan. 1960.
[171] A-12 Log, entries for 26 and 30 Jan. 1960, Lockheed ADP, Burbank, CA.
[172] Telex to Director, untitled, IN-23076, CIA, Washington, DC, 13 April 1960.
[173] Memo from Dan to Herb, DPD-5966-60, CIA, Washington, DC, 3 Aug. 1960.
[174] Memo from Paul to Dan, DPD-6971-60, CIA, Washington, DC, 12 Sept. 1960.
[175] Telex to Director, CIA, Washington, DC, 1 Feb. 1961.
[176] Shibata, H. H., "Proposed High Speed Wind Tunnel Tests of an Inlet Model 204 in the 8 × 7 Unitary Plan Wind Tunnel of the NASA at the Ames Research Center. Phase II," LAL 455-IIP, Lockheed ADP, Burbank, CA, 3 June 1960.
[177] Hill, Donald K., "Proposed High Speed Wind Tunnel Tests of an Inlet Model 204 in the Unitary-Plan Wind Tunnels of the NASA at the Ames Research Center," LAL 455-IVP, Lockheed ADP, Burbank, CA, 7 Dec. 1960.
[178] Hill, Donald K., "Proposed High Speed Wind Tunnel Tests of an Inlet Model 204 in the 8 × 7 Unitary-Plan Wind Tunnels of the NASA at the Ames Research Center," LAL 455-VIIP, Lockheed ADP, Burbank, CA, 1 Aug. 1961.
[179] Abernethy, R. B., "More Never Told Tales of Pratt & Whitney," presentation at J-58 Eagles Reunion, Los Angeles, CA, 26 March 2004.
[180] Goodall, James, and Jay Miller, *Lockheed's SR-71 "Blackbird" Family: A-12, F-12, M-21, D-21, SR-71*, Midland Publishing, Hinckley, England, 2002, pg. 91.
[181] Eldridge, Arnold L., "Object Camouflage Method and Apparatus," U.S. Patent 3,127,608, filed 6 Aug. 1956.
[182] Bissell, R. M., Jr., "RAINBOW Investigation," SAPC-10098, Lockheed ADP, Burbank, CA, 29 October 1956.
[183] Engineering and Analysis Division, "Trip Report to [redacted] New York," CIA, Washington, DC, 19 July 1963.
[184] [KEMPSTER Project] Final Report Volume II, Equipment Development Programs, Westinghouse Research Laboratories, Westinghouse Electric Corporation, Pittsburgh, Pennsylvania, 30 June 1965, 288 pp.

[185] Chief, Air Systems Division, OEL, "Request for Funds for Project KEMPSTER A, Phase II," CIA, Washington, DC, 15 Oct. 1965.
[186] Parangosky, J., "Funding Status of KEMPSTER A Program," document number [redacted]-2568-66, CIA, Washington, DC, 12 Aug. 1966.
[187] Poteat, S. E., "Stealth, Countermeasures, and ELINT, 1960–1975," *Studies in Intelligence*, Vol. 42, No. 1, 1998, CIA, Langley, VA, pp. 51–59.
[188] Vovodich, M., speech to Roadrunners Internationale, Las Vegas, Nevada, October 1995.
[189] McIninch, T. P. [John Parangosky], "The OXCART Story," *Studies in Intelligence*, Vol. 15, No. 1, Winter 1971, CIA, Langley, VA, pp. 1–34.
[190] "Stenotype Transcript of Press Conference," Press Conference No. 23, Press Stenotypists Association, Washington, DC, July 24, 1964.
[191] Mulready, D., *Advanced Engine Development at Pratt & Whitney*," Society of Automotive Engineers, Inc., Warrendale, PA, 2001, pp. 103–116.
[192] Bacalis, P. N., "Memorandum for Director of Reconnaissance, CIA: ISINGLASS," CIA, Langley, VA, 15 Aug. 1966.
[193] Whittenbury, J. R., "From ARCHANGEL to OXCART: Design Evolution of the Lockheed A-12, First of the Blackbirds," Lockheed Martin Aeronautics Company, Palmdale, CA, 1999.

BIBLIOGRAPHY

NOTES ON SOURCES

There are relatively few sources on the SUNTAN (CL-400) aircraft. The primary source is Sloop's NASA report. Dick Mulready devoted a chapter of his book to the Model 304 engine. Goodall and Miller also provide a good overview. Citations for these three books are in the following.

The CIA documents cited in this work were obtained from the CIA Records Search Tool (CREST) computer system at the National Archives II Library, College Park, Maryland. Cited documents can be looked up in the GUSTO documents index in the attached CD-ROM to obtain their "CIA RDP number," which in turn can be used to access the CREST system. A CREST finding aid is available online at http://www.foia.cia.gov/search_archive.asp. This tool can give the titles and dates of documents in CREST, but not the documents themselves.

The Jay Miller Aviation History Collection has a number of documents pertaining to Convair's Project FISH, as well as much of the source material used in the preparation of Miller's 1993 official history of the Skunk Works. The collection is located at the Aerospace Library in the Aerospace Education Center Complex, Little Rock, Arkansas.

DRAWINGS

Baldwin, E. P., U-2 GA Drawing, Revision A, 29 Jan. 1955, Lockheed ADP, Burbank, CA.
Baldwin, E. P., B-2 GA Drawing, 3 Dec. 1957, Lockheed ADP, Burbank, CA.
Baldwin, E. P., All Metal Airplane GA Drawing, 9 Dec. 1957, Lockheed ADP, Burbank, CA.
Baldwin, E. P., G2 GA Drawing, 5 Feb. 1958, Lockheed ADP, Burbank, CA.
Baldwin, E. P., A-1 GA Drawing, 3 July 1958, Lockheed ADP, Burbank, CA.
Baldwin, E. P., Peterbilt GA Drawing, 22 Aug. 1958, Lockheed ADP, Burbank, CA.
Baldwin, E. P., A-2 GA Drawing, 3 Sept. 1958, Lockheed ADP, Burbank, CA.
Baldwin, E. P., Cherub #1 GA Drawing, 13 Oct. 1958, Lockheed ADP, Burbank, CA.
Baldwin, E. P., A-3 GA Drawing, 30 Oct. 1958, Lockheed ADP, Burbank, CA.
Baldwin, E. P., A-4 GA Drawing, 11 Dec. 1958, Lockheed ADP, Burbank, CA.
Baldwin, E. P., A-7-1 GA Drawing, 6 Jan. 1959, Lockheed ADP, Burbank, CA.

Baldwin, E. P., A-7-2 GA Drawing, 7 Jan. 1959, Lockheed ADP, Burbank, CA.
Baldwin, E. P., A-7-3 GA Drawing, 15 Jan. 1959, Lockheed ADP, Burbank, CA.
Baldwin, E. P., A-10 GA Drawing, undated, Lockheed ADP, Burbank, CA.
Baldwin, E. P., A-11 GA Drawing, 12 March 1959, Lockheed ADP, Burbank, CA.
Baldwin, E. P., A-11A GA Drawing, 20 March 1959, Lockheed ADP, Burbank, CA.
Baldwin, E. P., R-12 GA Drawing, 4 Dec. 1962, Lockheed ADP, Burbank, CA.
Baldwin, E. P., TR-12 GA Drawing, 18 Feb. 1964, Lockheed ADP, Burbank, CA.
Baldwin, E. P., AF-112 GA Drawing, 5 March 1964, Lockheed ADP, Burbank, CA.
Baldwin, E. P., A-12CB GA Drawing, 24 Feb. 1965, Lockheed ADP, Burbank, CA.
Baldwin, E. P., B-71 GA Drawing, 19 March 1959, Lockheed ADP, Burbank, CA.
Combs, Henry G., A-3 GA Drawing, 14 Oct. 1958, Lockheed ADP, Burbank, CA.
Combs, Henry G., B-58 Launched Vehicle GA Drawing, 12 Dec. 1958, Lockheed ADP, Burbank, CA.
Combs, Henry G., B-58 Launched Vehicle on B-58 GA Drawing, 12 Dec. 1958, Lockheed ADP, Burbank, CA.
Johnson, C. L., Archangel first concept drawing, 26 June 1958, Lockheed ADP, Burbank, CA.
Nystrom, H., A-12 GA Drawing, 1959, Lockheed ADP, Burbank, CA.
Taylor, W., AF-12 GA Drawing, 23 May 1961, Lockheed ADP, Burbank, CA.
Unidentified author, A-12 GA Drawing, July 1959, Lockheed ADP, Burbank, CA.
Zuck, Daniel R., High-wing Ram Jet Kite GA Drawing, 23 Aug. 1958, Lockheed ADP, Burbank, CA.
Zuck, Daniel R., Mid-wing Ram Jet Kite GA Drawing, 25 Aug. 1958, Lockheed ADP, Burbank, CA.
Zuck, Daniel R., A-3 GA Drawing, Oct. 1958, Lockheed ADP, Burbank, CA.
Zuck, Daniel R., Cherub #2 GA Drawing, 10 Oct. 1958, Lockheed ADP, Burbank, CA.
Zuck, Daniel R., Arrow I GA Drawing, 3 Dec. 1958, Lockheed ADP, Burbank, CA.
Zuck, Daniel R., A-5 GA Drawing, 5 Dec. 1958, Lockheed ADP, Burbank, CA.
Zuck, Daniel R., A-6-5 GA Drawing, 9 Jan. 1958, Lockheed ADP, Burbank, CA.
Zuck, Daniel R., A-6-6 GA Drawing, 20 Jan. 1959, Lockheed ADP, Burbank, CA.
Zuck, Daniel R., A-6-9 GA Drawing, 30 Jan. 1959, Lockheed ADP, Burbank, CA.

Audio Recordings

Baldwin, E. P., oral history, unpublished audio recording.

Articles

Baldwin, E. P., "U-2 General Arrangement," unpublished memo, undated.
Baldwin, R. E., unpublished transcript of tour of SR-71 by Ed Baldwin and Lou Wilson, Beale Air Force Base, California, 30 Aug. 1989.
Brown, A., "Fundamentals of Stealth Design," *Lockheed Horizons*, Issue 31, Aug. 1992, pp. 6–12.
Brown, W. H., "J58/SR-71 Propulsion Integration or the Great Adventure into the Technical Unknown," *Lockheed Horizons*, no. 9, Winter 1981/82, pp. 7–13. Reprinted in *Studies in Intelligence*, Summer 1982, pp. 15–23.
Hall, R. C., "Postwar Strategic Reconnaissance and the Genesis of Corona," Chapter 4 in *Eye in the Sky: The Story of the Corona Spy Satellites*, ed. by Dwayne A. Day, John M. Logsdon, and Brian Latell, Smithsonian Institution, 1998, 303 pp.

Johnson, Clarence L., "Development of the Lockheed SR-71 Blackbird," *Lockheed Horizons*, no. 9, Winter 1981/82, pp. 2–7 and 13–18. Reprinted in *Studies in Intelligence*, Summer 1982, pp. 3–14.

Johnson, C. L., "Reduction of Radar Cross Section of Large High Altitude Aircraft," Proceedings of the 1975 Radar Camouflage Symposium, Oct. 1975, Wright Patterson Air Force Base, OH.

Johnson, C. L., "Some Development Aspects of the YF-12A Interceptor Aircraft," *Proceedings of AIAA 69-757*, July 14–16, 1969, Los Angeles, CA.

Poteat, S. E., "Stealth, Countermeasures, and ELINT, 1960–1975," *Studies in Intelligence*, Vol. 42, No. 1, 1998, pp. 51–59.

Rich, B. R., "F-12 Series Aircraft Aerodynamic and Thermodynamic Design in Retrospect," *Journal of Aircraft*, Vol. 11, No. 7, July 1974, pp. 401–406.

Rodgers, F. A, unpublished memoir, 3 Feb. 1995.

Taylor, H. H., unpublished memoir, 2002.

REPORTS AND PRESENTATIONS

Abernethy, R. B., "More Never Told Tales of Pratt & Whitney," J-58 Eagles Reunion, Los Angeles, CA, 26 March 2004; also available at http://www.bobabernethy.com/blackbirds_presentation.htm

Johnson, C. L., Archangel project design notebook, Lockheed ADP, Burbank, CA, 21 April–29 Dec. 1958.

Johnson, C. L., *A-12 Log*, Lockheed Martin Aeronautics Company, Palmdale, CA, via Jay Miller Collection and the Aerospace Education Center.

Johnson, C. L., *Design Study: Archangel Aircraft*, SP-101, Lockheed Aircraft Corporation Advanced Development Projects, Burbank, California, 11 Sept. 1958, 14 pp.

Johnson, C. L., *History of the OXCART Program*, SP-1362, Lockheed Aircraft Corporation Advanced Development Projects, Burbank, California, 1 July 1968, 25 pp.

Johnson, C. L., *Proposal for a Lightweight Reconnaissance Aircraft*, SP-108, Lockheed Aircraft Corporation Advanced Development Projects, Burbank, California, 10 Nov. 1958, 24 pp.

Johnson, C. L., excerpts from *U-2 Log*, Lockheed Martin Aeronautics Company, Palmdale, CA, 16 Aug. 1956–Feb. 1958, via Chris Pocock.

[KEMPSTER Project] Final Report Volume II, Equipment Development Programs, Westinghouse Research Laboratories, Westinghouse Electric Corporation, Pittsburgh, Pennsylvania, 30 June 1965, 288 pp.

Project Fish Status Review, PF-0-101M, Convair, Fort Worth, TX, 9 June 1959, 32 pp.

McIninch, T. P. [John Parangosky], "The OXCART Story," *Studies in Intelligence*, Vol. 15, No. 1, Winter 1971, pp. 1–34.

Nelson, N. E., and Reichert, J. B., "Proposal for a High Altitude Reconnaissance Aircraft," Report No. DS-33-01, 22 June 1966, Lockheed ADP, Burbank, CA, via Norman Nelson.

Siegel, K. M., Alperin, H. A., Bonkowski, R. R., Crispin, J. W., Maffett, A. L., Schensted, C. E., and Schensted, I. V., "Bistatic Radar Cross Sections of Surfaces of Revolution," Willow Run Research Center, University of Michigan, Ann Arbor, Michigan, April 9, 1954.

Sloop, J. L., *Hydrogen as a Propulsion Fuel, 1945–1959*, SP-4404, NASA History Series, National Aeronautics and Space Administration, Washington, DC, 1978.

Super Hustler: A New Approach to the Manned Strategic Bombing-Reconnaissance Problem, FZM-1200-20, Convair, Fort Worth, TX, 26 May 1958, 279 pp.

Super Hustler SRD-17 TAC Bomber Studies, FZM-1556B, Convair, Fort Worth, TX, 27 April 1960, 20 pp.

Taylor, W., "RB-12 Proposal," SP-229, Lockheed Aircraft Corporation Advanced Development Projects, Burbank, California, via the Jay Miller Collection and the Aerospace Education Center.

Vojvodich, M., speech to Roadrunners Internationale, Las Vegas, NV, Oct. 1995. Transcribed by Joseph Donoghue.

Wheelon, A. D., speech to Roadrunners Internationale, Las Vegas, NV, 2 Oct. 2003.

Whittenbury, J. R., "From ARCHANGEL to OXCART: Design Evolution of the Lockheed A-12, First of the Blackbirds," Lockheed Martin Aeronautics Company, Palmdale, CA, 1999, Public Information Release Authorization #AER200305006.

PATENTS

Abernethy, R. A., "Recover Bleed Air Turbojet," U.S. Patent 3,344,606, filed 27 Sept. 1961, patented 3 Oct. 1967.

Campbell, D. H., "Supersonic Inlet for Jet Engines," U.S. Patent 3,477,455, filed 15 Oct. 1965, patented 11 Nov. 1969.

Eldredge, A. L., "Object Camouflage Method and Apparatus," U.S. Patent 3,127,608, filed 6 Aug. 1956, patented 31 March 1964.

Gothe, A., "Method of Eliminating Reradiation," U.S. Patent 2,103,358, filed 28 Dec. 1937, patented 28 Dec. 1937.

Salisbury, W. W., "Absorbent Body for Electromagnetic Waves," U.S. Patent 2,599,944, filed 11 May 1943, patented 10 June 1952.

Skellett, A. M., "Shield for Electromagnetic Radiations," U.S. Patent 2,828,484, filed 3 June 1947, patented 25 March 1958.

BOOKS

Bissell, R. M., Jr., *Reflections of a Cold Warrior: From Yalta to the Bay of Pigs*, Yale University Press, 1996.

Goodall, J., and Miller, J., *Lockheed's SR-71 'Blackbird' Family: A-12, F-12, M-21, D-21, SR-71*, Midland Publishing, Hinckley, England, 2002.

Johnson, C. L., *Kelly: More than My Share of It All*, Smithsonian Books, 1985.

Mulready, R., *Advanced Engine Development at Pratt & Whitney*, SAE International, Warrendale, PA, 2001.

Price, A., *The History of U.S. Electronic Warfare, Volume II*, The Association of Old Crows, Alexandria, VA, 1989.

Rich, B., and Janos, M., *Skunk Works*, Back Bay Books, Newport Beach, CA, 1996.

INTERVIEWS BY AUTHOR

Bazemore, Thomas C., Santa Barbara, CA, 13 Nov. 2004.
Combs, Henry G., Santa Clarita, CA, 19 April 2003.
Eastham, James D., Rancho Palos Verdes, CA, 21 Dec. 2002.
Gath, Elmer, Mission Viejo, CA, 28 March 2004.
Geary, Leo P., Denver, CO, 10 July 2005.
Kelder, Sam, Cottage Grove, OR, 12 Jan. 2003.
Lovick, Ed, Northridge, CA, 4 Feb. 2006.
Lovick, Ed, Northridge, CA, 8 March 2008.
Murray, Frank, Las Vegas, NV, 1 Oct. 2007.
Naka, F. Robert, Concord, MA, 16 Sept. 2004.

Nelson, Norman E., Palos Verdes Peninsula, CA, 2 March 2002.
Pendleton, Wayne E., Torrance, CA, 9 Oct. 2005.
Taylor, Norman H., Topsfield, MA, 1 Dec. 2003.
Taylor, William, Hollywood, CA, 17 Nov. 2003.
VanDerZee, Charles, Cottage Grove, OR, 12 Jan. 2003.
Schwarzkopf, Daniel, and Brint Ferguson, Stow, MA, 30 Nov. 2003.
Widmer, Robert, Fort Worth, TX, 13 March 2003.
Widmer, Robert, and Randy Kent, Fort Worth, TX, 7 Nov. 2003.

TELEPHONE INTERVIEWS BY AUTHOR

Abernethy, Robert B., 12 June 2004.
Bissell, William, 13 June 2005.
Bissell, William, 15 June 2005.
Boyd, Robert, 9 Feb. 2008.
Brown, Alan, 15 Aug. 2005.
Bullock, Frank, 23 Jan. 2008.
Butman, Robert, 2 Dec. 2003.
Combs, Henry, 3 June 2001.
Geary, Leo P., 31 July 2002.
Geary, Leo P., 21 Aug. 2002.
Geary, Leo P., 24 May 2003.
Leghorn, Richard, 2 July 2004.
Warwick, Thomas, 16 June 2004.
Warwick, Thomas, 18 June 2004.
Warwick, Thomas, 22 Oct. 2004.
Warwick, Thomas, 23 April 2008.
Wheelon, Albert D., 3 April 2002.

ELECTRONIC MAIL TO AUTHOR

Abernethy, R. A., 4 Feb. 2008.
Bell, G., 13 Feb. 2007.
Boyd, B., 13 Feb. 2008.
Donoghue, J., 9 Oct. 2007.
Lovick, E., 1 March 2008.
Pendleton, W. E., 21 Feb. 2008.
Schwarzkopf, D., 22 Oct. 2007.
Schwarzkopf, D., 16 April 2008.
Taylor, N. H., 25 Sept. 2003.
Warwick, T., 23 June 2004.
Warwick, T., 24 June 2004.
Warwick, T., 26 June 2004.
Warwick, T., 17 July 2004.
Warwick, T., 18 July 2004.

INDEX

3M (Minnesota Mining and Manufacturing), 50
3X1000 project, 17
6901st Special Communications Group, 36

A-3 design, 119, 122
A-4 design, 129–133
A-5 design, 133–134
A-6 design, 134–137
A-6-5 design, 134–135
A-6-6 design, 135
A-6-9 design, 135, 136
A-7 design, 139–140
A-8 design, 139, 140
A-9 design, 139, 140
A-10 design, 155
A-11 design, 155–157, 159, 161, 177
A-12 design, 65, 172–175, 177, 180, 187, 219
  with 204 inlet, 219
  airflow, 217–218
  bomber variant, 242
  cost of, 181
  design revision, 208–210
  dish antenna and, 213–215
  first mission, 235–236
  full-scale RCS model of, 188
  inlet guide vanes, 223–224
  ionized exhaust, 203–204
  late changes, 211–212
  new mission, designing for, 221
  PDP-3, 210–211
  progress, 204–208
  radar-absorbent materials (RAM), 196–202
  testing of, 190–196
Abernethy, Robert B., 220

Advisory Panel. *See* Land Panel
Aero-Electronics Branch, 50
Aerojet, 130, 134
Aeronautic Research Institute (DVL), 17
Aeronautical Photographic Laboratory, 2
AFDAP. *See* Air Force's Development and Advanced Planning
Afghanistan, 36
Agency (CIA). *See* Central Intelligence Agency
Agency Reserve funds, 101
AGM-28 missile, 148
Air Council, 78, 95
Air Defense Command, 20
Air Force's Development and Advanced Planning (AFDAP), 3
Air Research and Development Command (ARDC), 46, 50, 71, 72, 74, 76
Air Technical Intelligence Center (ATIC), 53, 99, 175, 178
Alexander, N. M., 90
All-Metal Airplane design, 57, 67
Allmon, Carl, 6
American intercontinental ballistic missile (American ICBM), 142
Ames Research Center, 206, 218
Ampex tape drive, 210
AN/ASG-18 radar, 237
Anderson, Samuel, 46
Anechoic chamber, 22, 23, 42, 55, 61
Angel (U-2), 82
Applied Physics Laboratory of Johns Hopkins University, 71
Appold, Norman, 74, 184
Aral Sea, 37
Archangel I design, 95–101
Archangel II design, 106–108, 109, 112

ARDC. *See* Air Research and Development Command
Arrow I design, 137, 139
Asbestos, 65
Ash, Michael, 22
ATIC. *See* Air Technical Intelligence Center
Autonetics, 148

B-2 design (Lockheed), 55
B-2 Spirit stealth bomber, 244
B-47, 176, 180
B-52, 72
B-58, 73, 126, 152
B-58 launched vehicle design, 137, 138
B-58A carrier, 150
B-58A engines, 179
B-58B, 160, 161, 162, 179
B-70, 180
Baker, Jim, 2
Bakersfield, 23
Baldwin, Edward, 6, 7, 56, 57, 58, 59, 60, 97, 98, 103, 105, 107, 115, 116, 133, 136, 139, 156, 157, 197, 212
BARLOCK radar, 175
Bathtub, 42–43
Batplane design, 63–64
Bazemore, Thomas C., 1, 14, 31
Beacon Hill Study, 3
Bell Laboratories, 2, 26
Bendix, 225
Bevins, Fred, 41
Big Boy Rigging Company, 34
Big rumbling cart, 22
Binney Street, 41, 42, 43
Bissell, Bill, 7
Bissell, Richard M., Jr., 5, 8, 9, 11, 13, 32, 36, 37, 39, 40, 46–50, 51, 53, 55, 63, 70, 81–82, 85, 86, 93, 94, 99–101, 112, 113, 122, 123, 126, 127, 129, 137, 140, 141, 144–145, 146, 161, 162, 163, 167, 178, 180, 184, 185, 188, 190, 203, 207, 208, 230, 243
Bjorksten Research Laboratories, 38
BLACK SHIELD, 235
Blip-Scan Study, 82–84
Boehme, Dick, 6, 95, 97, 134
Boeing Aircraft, 77, 78, 86

Borane fuel, 106, 107, 119
Boron (CA), 20
Boston University Optical Research Laboratory (BUORL), 3
Boyd, Bob, 225
Bradley, Mark, 176
British Miles M.52, 105
Brooklyn Polytechnic University, 38
Brown, Charles F., 41
Brown, Bill, 220, 222, 230
Bunce, Donald, 233
BUORL. *See* Boston University Optical Research Laboratory
Burbank, 5, 18, 23, 63
Bureau of the Budget, 101
Burke, William, 113, 159, 185, 186
Butman, Bob, 15, 37, 41
Bykhov, 12

Cabell, Charles P., 37, 180
California Institute of Technology (CalTech), 8, 85
Cambridge Research Center, 13
Campbell, David, 217, 218
Carswell Air Force Base, 126
CASINO. *See* Computer able to select internal orders
Cat and Mouse Study, 159
CATNAP, 234
Central Intelligence Agency, 1, 2, 35, 143
  Office of Security Support, 143
CHAMPION, 103–106
Cherbonneaux, Jim, 36
Cherub designs, 115–116, 117
China, 36, 53
Chine, 64–65
CL-282, 3, 4, 5, 7
CL-325 design, 71–74
CL-400 design, 74–77
CMR, 36
Coar, Dick, 221
Combs, Henry, 7, 74, 95, 97, 115, 137, 156–157, 187
Computer able to select internal orders (CASINO), 211
Concord, MA, 24, 25, 27
Conductor, 225
Contract BE-2022, 34
Contract HL-4646, 146
Contract TE-2191, 189

INDEX

Convair, 85, 94, 122–123, 125, 126, 143, 144, 145, 146, 147, 149, 160, 161, 162, 163, 164, 166, 167, 168, 172, 176, 181, 184
  B-58, 226
  ramping down, 183–185
  WO 540, 243
  XP-92, 104
Copacabana, 25
Corderman, Chuck, 41, 210
Corona spy satellite project, 141
Cotter, Norman, 222
Covered Wagons, 36
Cuba, 235
Cunningham, James A., 8

D. S. Kennedy Company, 213
D-21 drone, 241, 242, 244
Daily mission report (DMR), 36
Damaskos, Nick, 231
Daniel, Russ, 6
DCI. *See* Director of Central Intelligence
DEC. *See* Digital Equipment Corporation
Definitive Contract HL-4646, 146
Demler, Marvin, 46, 50, 162
Department of Defense (DoD), 2
Development Projects Division (DPD), 123
DEW. *See* Distant Early Warning Line
Diffuser, 217
Digital Equipment Corporation (DEC), 210
Dipole antenna, 28, 49
Director of Central Intelligence (DCI), 11
Dirty birds, 28, 36, 62
Distant Early Warning (DEW) Line, 13, 213
DMR. *See* Daily mission report
DoD. *See* Department of Defense
Dolson, Vincent, 90
Donovan, Allen, 3
Dow Chemical, 211
DPD. *See* Development Projects Division
Duckett, Carl, 233
Dulles, Allen, 4, 5, 11, 13, 53, 101, 180, 181, 208
Dulles, John Foster, 11
Duncan, James, 76
DVL. *See* Aeronautic Research Institute

East Hartford, 77
Eastham, James D., 212, 237, 241
Eastman Kodak, 41
ECM. *See* Electronic counter measures

Edens, Eugene, 37
Edgerton, Germeshausen, & Grier (EG&G), 34, 172, 211, 213
Edwards Air Force Base, 20, 27
Eglin Air Force Base, 28
EGT. *See* Exhaust gas temperature
Eisenhower and flying saucer, 141–142
Eisenhower, Dwight D., 2, 5, 11, 123, 181
Eldredge, Arnold, 18
Electronic counter measures (ECM), 37–38
Electronic intelligence (ELINT), 11
Ellis, Stan, 228
Emerson and Cuming, 38, 61
English, Leroy, 196
Esenwein, August, 161
Esmeier, Edward, 221
Evaluation Studies of Inflatable High-Altitude Aircraft, 105–106
Exhaust gas temperature (EGT), 230

F8U Crusader, 80
F-12, 237
F-94, 188, 195, 205
F-104, 4, 6, 7, 80
F-108, 80, 99, 180
F-111, 94
F-117, 59
Fairchild, 50
FAN SONG radar, 24
FBW. *See* Fly-by-wire system
Ferguson, Brint, 41, 43, 203, 214
Ferri, Antonio, 90
Fire control radar, 213
First Invisible Super Hustler. *See* FISH
First mission, 235–236
FISH
  B-58 carrier, 150, 151
  cost of, 181
  demise of, 161–162
  design, 90–94, 121–122, 124, 137–138, 159, 162, 163, 177, 179, 184, 185
  desk model, 148
  facilities, 143–147
  new models, 148–153
  production scale, 160
  refining, 143
  resurrection of, 176
  subsystems, 147–148
  turbojet installation, change in, 153
  wind-tunnel model, 149

Flickinger, Don, 46
Flight testing, 32–33
Fly-by-wire (FBW) system, 148
Flying saucer, 38, 64, 141
Fort Worth, 143
FORTRAN programming language, 226
Fuller, Dick, 96, 97, 107, 172
Fur burger, 105
FX-114 engine prototype, 220

GAR-9 missile, 237, 238
Gardner, Trevor, 3, 72
Garrett Corporation, 71, 72
Gath, Elmer, 6, 97
Gavette, Leon, 190
GCI. *See* Ground-controlled-intercept operations
Geary, Leo P., 8, 9, 63, 86, 127, 187
GEBO. *See* Generalized Bomber
General Applied Science Laboratories, Inc., 90
General Arrangement #2 (GA#2) design, 59, 60
General Dynamics, 85
General Electric, 18, 74, 88
Generalized Bomber (GEBO), 86
George, Melvin F., 22, 31, 203
Gibbs, Jack A., 8, 21, 36, 50, 52, 69, 81
Glomar Explorer, 187–188
Goodpaster, Andrew, 11, 12, 180
Goodyear Corporation, 70, 103
    Inflatoplane project, 106
Gordon, William, 221
Gore, Bill, 131
Graphite, 16, 24, 65, 92
Grissom, Virgil I., 38
Grogan, C. G., 90
Ground-controlled-intercept (GCI) operations, 45
GUS-0285 memo, 159
GUSTO, 185
GUSTO 2 design, 64–69
GUSTO 2A, 112
GUSTO supersonic designs, 78–82

Hairflex, 202
Hamilton Standard Corporation, 225, 229
HAVE BLUE, 17, 59
Heal, Merv, 97
HEF. *See* High-energy fuel
Herron, James M., 22

Hibbard, Hall, 5
High-energy fuel (HEF), 92, 221
Hippert, Bob, 112
Ho229, 17
Holloway, Marshall, 13, 14, 31
Holzapple, Joseph R., 162
Horner, Richard, 69
Horten Ho229 flying wing, 17
Horten, Reimar, 17
Hughes Aircraft Corporation, 237
Hycon, 244
Hydrogen peroxide fuel, 130, 134

IBM 709 computer, 226
Iconel-X, 153
Identification friend or foe (IFF), 21
IGVs. *See* Inlet guide vanes
Indian Springs Air Force Base, 31, 63, 94, 170
Inertial navigation, 124, 148
Inflatoplane, 103, 106
Infrared, 58, 124, 159, 164, 178, 190
Inlet, 36, 56, 59
Inlet guide vanes (IGVs), 223–226
Internal research and development (IRAD) funds, 122
Iron Maiden, 84, 100
Irvine, Clarence, 74
Itek Corporation, 41

J57 engine, 7, 8, 179
J58 (JT9) engine, 77, 79, 96, 97, 107, 132, 134, 164, 217
    problems of, 179–180
    redesign, solving problems, 222–223
    six bypass tubes, 224
    testing, 228–230
J75 engine, 104, 112, 228, 229
J79 engine, 64, 160
J85 engine, 88, 90, 93
J93 engine, 79, 155, 220
Japan, 1, 36
Jaumann, Johannes, 17
Jet Propulsion Laboratory, 149
Johnson, Clarence L., 5, 8, 9, 13, 18, 26, 30, 40, 56, 60, 71, 74, 76, 81, 90, 95, 103, 104, 105, 106–112, 113, 114, 115, 121, 122, 129, 130–131, 132, 134, 137, 138, 141, 159, 163, 187, 188, 189, 190, 197, 203, 205, 206, 207, 208

Johnson, Lyndon, 240, 241
Jones, L. K., 37
JP-5 fuel, 130
JP-150, 107
JT9, 220
JT-12 engine, 93
JT-12As, 116, 134

Kadena, 235
Kapustin Yar, 3, 46
Katzenstein, Henry, 25
KC-135, 178–179
Kennedy, Joseph, 2
Kent, Randy, 185
Kiefer, Eugene, 37, 63, 81, 95, 121, 126, 127, 148, 163, 167, 168, 175, 176, 183, 200
Killian, James R., 2, 54, 122, 123, 142, 176, 180
KINGFISH design, 164–172, 177, 179, 181, 184
Kirby & McGuire, 34
Kirk, Donald R., 90
Kirkham, Ray, 63, 64
Kistiakowsky, George, 180
Klein, Joe, 41, 43, 203
Klinger, Bob, 28

L-182 design, 72
LaCroix, Benjamin, 231
Land Panel, 94, 112–115
Land panel review, 159
Land, Edwin H., 1, 2, 142, 208
Las Vegas, NV, 8, 21
Latmiral, Gaetano, 37
Lawson, Jay, 41, 210, 215
L-band, 49
Leghorn, Richard, 2, 3, 41
LeMay, Curtis, 161, 162
Leningrad, 12, 46
Lewis, C. H., 37, 50
Lincoln Laboratory at Massachusetts Institute of Technology, 1, 13, 15, 40, 83
Liquid hydrogen, 71, 72, 73, 76, 78, 95
Lockheed, 103, 122, 159, 160, 162, 163, 181, 182, 187, 189, 200, 203
  A-4, 129–133
  A-5, 133–134
  A-6, 134–137
  A-7, 139–140
  A-10, 155
  A-11, 155–157
  Advanced Development Projects, 1, 3, 5
  D-33, 243
  David Campbell, 218
  FISH, 137–138
  large airplanes, 155
  Mach 3 aircraft, 237
  stealthy designs, 129
  supersonic refueling, 140–141
Los Angeles, 18, 32
Lovick, Edward, 22, 23, 26, 27, 28, 40, 55, 61, 64, 97, 172, 189, 203
Lundahl, Arthur, 144

M-21, 241
MacDonald, Luther D., 22, 27, 68, 97, 134, 172, 189, 196
Mach 3 engine, 220
  bypass doors, 226
Magic T waveguide, 22
Mansville, John, 212
Marquardt Corporation, 90
Marquardt, Ray, 108, 146, 149
Martin, Edward, 97
Matteson, R. C., 90, 127
McCone, John, 235
McDonnell Aircraft, 244
McElroy, Neil, 142, 180
McMurray, Bill, 37
McNamara, Robert S., 241
McNarney, J. T., 161, 162
Metallic salts, 203, 204
Mice, 230
MiG Ye-2A, 45
MiG-17, 45
MiG-19, 45
Miles M.52, 105
Miller, Herbert, 12, 14, 18, 20, 41
Miller, Stewart, 26
Minnesota-Honeywell (M-H), 147
Minnesota Mining and Manufacturing. *See* 3M
Minority report, 175–176
Model 304 engine, 76, 78
Moscow, 46
Mylar, 42, 49

NACA. *See* National Advisory Committee for Aeronautics

Nacelle, 217
Naka, F. Robert, 1, 13, 24, 25, 31
Naka, Patricia, 25
Narmco, 50
NASA. *See* National Aeronautics and Space Administration
National Advisory Committee for Aeronautics (NACA), 52
National Aeronautics and Space Administration (NASA), 75, 76
　Ames Research Center, 218
National Reconnaissance Office (NRO), 233, 244
National Security Agency (NSA), 36, 37
Naval Air Systems Command (NAVAIR), 225
Naval Research Laboratory (NRL), 38
Navy inflatable designs, 103–106
Nelson, Don, 107
Nethkin, Harley, 203
New reconnaissance system, 46–48
Nike radar, 214
Nonstealthy designs, 138–139
North American Aircraft, 77, 80
North Vietnam, 235, 236
November 1958 Land Panel review, 121
　backchannel, 127
　comparison, 124–125
　FISH, choosing, 121–122
　funding, 122–123
　further studies, 125–127
　requirements, 124
　second thoughts, 123
　White House approval, 123
NRL. *See* Naval Research Laboratory
NRO. *See* National Reconnaissance Office
NSA. *See* National Security Agency
Nuclear Testing Site, 8
Nunziato, Ralph J., 37, 46
Nystrom, Herb, 209

Oblique shock, 217
Occupational Safety and Health Administration (OSHA), 212
Office for Scientific Intelligence, 3, 175
Office for Special Activities (OSA), 185, 233
Office of Scientific Intelligence, 53
Office of Security Support, 143
OSA. *See* Office for Special Activities

OSHA. *See* Occupational Safety and Health Administration
Owens-Corning, 53, 69, 70
OXCART, 185–186, 187
The OXCART Story, 236

PALLADIUM, 234
Paradise Ranch, 8
Parangosky, John, 186, 188, 203
Pascal, Don, 221
PASSPORT VISA, 38, 61
PDP-3 computer, 210–211
Pendleton, Wayne, 206, 214, 215
Pentaborane, 107
Peterbilt design, 104
Plan position indicator (PPI), 20
Plastics, 69–70
Pocock, Chris, 3
Polaroid Corporation, 1, 2, 13, 38
Pole shield, 189, 206
Poteat, Gene, 234, 235
Power, Thomas, 74
PPI. *See* Plan position indicator
Pratt & Whitney, 76, 203, 207
Pratt & Whitney Florida Research and Development Center (FRDC), 225
Pratt, Perry, 109, 131
President's Board of Consultants on Foreign Intelligence (foreign Intelligence Board), 48
Project AQUATONE, 5, 11
Project CHALICE, 101
Project CHAMPION, 94
Project Charles, 13
Project CORONA, 100
Project GUSTO, 60–61, 78, 80, 100, 134, 143
Project IDIOM, 94
Project ISINGLASS, 244
Project KEDLOCK, 237
Project KEMPSTER, 231–236
Project OARFISH, 5
Project OXCART, 181, 185–186, 187, 207, 224, 227, 230, 234, 235
Project RAINBOW, 11, 15, 34, 45, 47, 71
Project SENIOR BOWL, 241
Project SUNTAN, 74–77, 78
Project TAGBOARD, 241
Proposal: A-11, 157

INDEX

Propulsion, 217
Purcell, Edward, 2, 14, 18, 49, 142, 208
Putt, Donald, 74
Pyroceram, 92

Quality ELINT, 235
Quarles, Donald, 46, 142

Radar absorbent material (RAM), 22, 23, 42, 53, 61, 196–202
Radar Camouflage Symposium, 66
Radar cross section (RCS), 16, 17, 83, 86
RADC. *See* Rome Air Development Center
Radiation Laboratory, 13, 14, 17
Radio Corporation of America (RCA), 21
Rae, Randolph, 71
Rainbow, 11
Rainbow phase II, 45
RAM. *See* Radar absorbent material
Ram Jet Kite three view drawings, 105
Ramjet, 86, 88, 89, 92, 94, 99
Ramo-Wooldridge, 35, 95
Ranch, The, 8, 20, 32
Rawson, Edward, 14, 20, 21, 26, 41, 43, 203, 210
RCA. *See* Radio Corporation of America
RCS model, 17, 22, 38, 40, 113, 114, 125, 162–163, 170
Reber, James, 36, 37
Reedy, Perry M., 22
Reichert, James, 243
Rene-41 steel alloys, 153
Rensselaer Polytechnic Institute, 85
Rex engine, 71–74
Rhombic antenna, 26
Rich, Benjamin, 107, 156
RJ-59 ramjet, 86
RL-10 engine, 78
Roadrunners Internationale, 196, 224, 230, 235
ROCK CAKE radar, 45
Rodgers Effect, 84
Rodgers, Franklin A., 1, 13, 15, 18, 19, 38, 41, 42, 121, 143, 187, 200
Rome Air Development Center (RADC), 123
RS-71. *See* SR-71

SAGE. *See* Semi-Automatic Air Ground Environment

Salisbury screen, 16, 17, 23, 24, 28
Salisbury, Winfield, 16
S-band, 11, 24, 30, 49
Schalk, Louis, 237
Schenk, Frederick, 230
Schornsteinfeger, 17
Schreiber (Convair engineer), 90
Schriever, Bernard, 3
Schwarzkopf, Daniel, 15, 19, 20, 30, 40, 203
Scientific Engineering Institute, 40–43, 61, 62, 63, 86, 90, 175, 197
Scoville, Herbert, 45, 46
SCR-584 radar, 20, 21
SD-3 radar, 21
Seaberg, John D., 112
Secord, C. L., 90
SEI. *See* Scientific Engineering Institute
Semi-Automatic Air Ground Environment (SAGE), 13
Sens, Bill, 132
SERN. *See* Single expansion ramp nozzle
Seybold, R. C., 161
SF-1 fuel. *See* Liquid hydrogen
Shock trap, 217
Shock wave, 89, 141, 174, 217, 218, 227, 229
Siegel, Kip, 17
Sieker, Bob, 33
Single expansion ramp nozzle (SERN), 164, 183
Skunk Works, 6, 7, 40, 85
Slater, Hugh, 236
Slow wave, 26, 27, 28
Small-scale RCS model, 155
Solis, A. E., 90
Sonic boom, 126, 159, 178
Specular reflection, 48
Spike, 219
Squid structure, 49
SR-71, 240–241
SRJ-54 ramjet, 109
Stalingrad, 37
Stever, Guyford, 208
STR-12 engine, 77
Strategic Air Command, 53
Strong, Philip, 3
Su-17, 45
Summers Gyroscope, 71, 72
Summers, Thomas, 71, 72, 85
SUNTAN, 74–77

Super Hustler, 86–90
Super Performance Rocket engine, 130
Supersonic refueling, 140–141
Surface-to-air missile, 46, 53
Swamp Monster, 77
Swofford, R. P., 112
System 5, 35–36

T-33, 28, 38, 61
TACIT BLUE, 59
TALL KING radar, 234–235
Tapered paper (TP), 197
Tashkent, 45, 53
Task Change Record No. 2, 145
Taylor, Bill, 7
Taylor, Norman H., 30, 39, 40, 86
TD. *See* Teledeltos paper
Technological Capabilities Panel (TCP), 2
Teledeltos paper (TD), 39, 197
Telemetry system, 214
Temperature-sensitive paint, 218
Teterboro Airport, 85
Thermos, 24
Thomas, D., 127
Thor missile, 141
Titanium, 78, 85, 97
Token radar, 11, 53
TP. *See* Tapered paper
Trans-Siberian Railroad, 45
Trapeze, 25–28, 39, 48, 63
Tug aircraft, 103–104, 105
Tukey, John, 2
Turkey, 36
Twining, Nathan F., 37, 53, 142

U-2, 5–9, 11, 19–22, 55–57, 111–112, 118
 navigation systems, 124
 tow plane design, 112
U-3 design, 79, 83
U-boat, 17
Ultra high frequency (UHF) band, 190
Union of Soviet Socialist Republics (USSR), 2, 11, 12, 36
United Technologies, 225
University of Goettingen, 37–38
University of Michigan Radiation Laboratory, 17
University of Naples, 37
Unstart, 226
Ural Mountains, 12, 45

Valley, George, 37
Verdan MBL-9A computer, 148
Vietnam War, 235
Vojvodich, Mele, 235
von Kármán, Theodore, 90
Vought F8U-3 Crusader III fighter, 220

Wallpaper, 22–25, 33, 63
Waltham, MA, 25
Ward and Ward, 206
Warwick, Thomas, 225–226, 227, 230
Watertown Strip, 8
West Palm Beach, 221, 227
Westinghouse, 33–35, 231
Wheeless, Hewitt T., 162
Wheelon, Albert D., 230, 233, 234, 235
White House approval, 123
White, Gerald, 46, 50, 180
White, Thomas D., 127
Widmer, Robert H., 85, 86, 87, 90, 121, 127, 159, 161, 162, 183, 185
Wiele, Robert, 6
Williams, Harry, 226
Wilson, John, 241
Wilson, Lou, 197, 212
Wind tunnel, 98, 116, 177
Wires, 28–30
Wood, Homer J., 71
Work Order 540, 185
Wright Field Aircraft Lab, 2, 70
Wright Field Materials Lab, 69
Wright, Rufus, 38

X-7 ramjet, 106
XB-70 Valkyrie bomber, 220
X-band, 51, 52, 62
XF-104, 7
XLR129 engine, 244
XP-92, 104
XPJ-59 ramjet, 139

Yagi antenna, 21

Zenith Corporation, 50
ZI. *See* Zone of the interior
Zip, 221
Zone of the interior (ZI), 124, 156, 159
Zuck, Daniel R., 68, 97, 115, 117, 133, 135